Importing Technology
into Africa

D. Babatunde Thomas
with contributions by
Mira Wilkins
Walter A. Chudson
Theodore W. Schlie

The Praeger Special Studies program—utilizing the most modern and efficient book production techniques and a selective worldwide distribution network—makes available to the academic, government, and business communities significant, timely research in U.S. and international economic, social, and political development.

Importing Technology into Africa

Foreign Investment and the Supply of Technological Innovations

PRAEGER SPECIAL STUDIES IN INTERNATIONAL ECONOMICS AND DEVELOPMENT

Praeger Publishers New York Washington London

Library of Congress Cataloging in Publication Data

Thomas, D   Babtunde.
   Importing technology into Africa.

   (Praeger special studies in international economics
and development)
   Bibliography: p.
   Includes index.
   1. Technological innovations—Africa—Addresses,
essays, lectures. 2. Technology transfer—Addresses,
essay, lectures. 3. Investments, Foreign—Africa—
Addresses, essays, lectures. I. Title.
HC505.T4T56   332.6'73'096   74-11606
ISBN 0-275-05740-2

PRAEGER PUBLISHERS
111 Fourth Avenue, New York, N.Y. 10003, U.S.A.

Published in the United States of America in 1976
by Praeger Publishers, Inc.

Printed in the United States of America

To

Ralph W. , Mary Elizabeth, and Elizabeth Abimbola

The main issues surrounding the transfer of technological innovations from the advanced, highly industrialized countries (ACs)
to the developing countries (DCs), or what has become known con-
temporaneously as the "Third" and "Fourth World," have attained a
considerable degree of urgency since the early 1970s. These issues
have become urgent because of the continuing gravity of development
problems in the DCs, especially unemployment, underemployment,
and dismal changes in labor productivity that have continued to com-
pound the widening gap between this group of countries and the ACs,
even in the presence of major efforts toward rapid industrialization.

The heterogeneity of DCs compels the preceding distinction
between the Third and Fourth World. This distinction is used to
convey recent empirical observations suggesting that, though economic and social progress have been made in some DCs, expressly
in per capita income, overall income distribution, health, education,
and other indicators of economic and social welfare, these changes
have not been substantial. Furthermore, there are several other
countries where progress is not evident. Based on the above indicators, these countries may be worse off. The group of countries in
which no apparent progress has been made in the last decade constitutes the Fourth World and a majority of African countries belong in
this category. According to the report of the United Nations Committee for Development Planning, 16 of the 25 countries classified
as "least developed" in 1972 are in Africa, one is in Latin America,
and the other eight are in Asia and Oceania. The group of countries
that in the past decade has been experiencing definite, though minimal, educational, technological, and industrial development, and,
in effect, has begun manifest economic progress and improved social
welfare, constitutes the Third World. The most prominent countries in this latter group are Argentina, Brazil, Mexico, and Taiwan.
The emerging technological-industrial structure in these countries
already represents a major factor in their development experience.
Direct, private foreign investment--DFI--represents the primary
channel for the inflow of the technological input in these countries.
There is, however, no conclusive evidence that DFI has actively promoted their educational, technological, and industrial development.
Although there has been similar technological input in the poorer
countries constituting the Fourth World, its effects on their development experience have been less apparent and very few studies have
been undertaken to ascertain the underlying causes of the relative

ineffectiveness of imported technology in raising employment and promoting economic progress.

The object of this book is to examine the manner in which technological innovations are being exported from the industrial and technologically advanced countries to African countries. The scope has been limited primarily to the past and future role of DFI in the flow of technological innovations. Major attention has been given to the analysis of the relationship between the inflow of foreign technological innovations and long-term technological development objectives in the domestic economies of African countries. Although lack of appropriate data was anticipated, and this problem invariably obviated the empirical verification of most of the propositions advanced in the following chapters, major ideological and conceptual problems inherent in the subject matter also compelled the diffused contents.

The first chapter examines in a partly historical and partly analytical manner many of the conceptual problems of technology "transfer" and DFI. It focuses primarily on recent situations.

Chapter 2 is Mira Wilkins' contribution. It provides a broad historical background to the contemporary situation covered in Chapter 1, and sets the stage for the subsequent chapters. Chapter 3 examines public policies on foreign investment and delineates the importance of national science and technology policy for long-term technological development. The implications of foreign investment policy for the flow of technological innovations are also considered in this chapter, but this is somewhat circumscribed. Chapter 4 considers the proposition that the precedence of choice of techniques by choice of products in DFI decisions by multinational corporations creates divergence between the requirements of imported technology and domestic input structure in DCs. Other implications of such choice making for consistent and appropriate technology flows are also examined. Chapters 5, 6, and 7 are case studies on Kenya, Nigeria, Tanzania, and Uganda. These chapters represent extensively revised and edited versions of papers presented at the workshop on "Barriers to Effective International Transfer and Diffusion of Technology to African Countries," which I chaired at the African Studies Association Annual Conference in October 1973 at Syracuse, New York. Chapters 5 and 6 are contributions by Theodore Schlie and Walter Chudson respectively. Chapter 8 presents a summary of the volume. It draws some general conclusions from the preceding chapters and offers policy recommendations.

I have received assistance from many sources in organizing this volume. A modest financial support from Ryder International, the School of Technology at Florida International University, and the U.S. Agency for International Development made it possible to pay for the travel expenses of the participants at the workshop in

Syracuse. The three contributors took part in the discussions of some of the topics covered in the five chapters I wrote for this volume. They also made my editorial work on their contributions a pleasant experience. Dr. Wilkins also read and made helpful comments on some of the chapters. My wife, Mary Ann, and my colleagues Jay Mendell, Edward Kuhnel, and numerous others also contributed in matters of substance to make the preparation of this book a meaningful experience. This is very much appreciated. I also owe thanks to Terisa Turner for her contribution to a short paper that was my initial effort on Chapter 7.

Finally, I am deeply indebted to Suzanne Pfenninger, Gina Greenstein, and Brenda Gentile for their cheerful assistance in typing various drafts of this manuscript.

CONTENTS

LIST OF TABLES

LIST OF FIGURES

1

DIRECT FOREIGN INVESTMENT AND THE FLOW OF TECHNOLOGICAL KNOW-HOW: SOME CONCEPTUAL PROBLEMS

The economic and social changes brought about, for better or for worse, by direct, private foreign investment or DFI, in developing countries--DCs--throughout the world have been far reaching.

Direct private foreign investment, as distinguished from private foreign portfolio investment, generally entails more than a financial flow. It gives the investor direct control over the management and use of the capital flows in the receiving enterprise, usually a wholly or partly owned subsidiary, affiliate, or branch. It therefore represents an extension of the foreign enterprise. Portfolio investment generally entails financial flows and ordinarily does not involve direct control. Although both are treated simply as private capital flows in balance-of-payments accounts and in the overall measurements of international capital movements, the focus in this book is DFI in view of its relative reliability as an index of the economic prospects in the recipient country, and its usual attendant organizational and technological knowledge, which are germane to an inquiry into the long-term technological development in DCs. DFI in the subsidiaries, affiliates, and branches of, principally, European and U.S. corporations is a dominant factor in the daily conduct of economic activity in virtually all African countries.*

Theories that attempt to explain the reasons for international flows of private capital investment have focused primarily on differences in the rates of interest and, consequently, on the relative profitability of investment in alternative markets. These common postulates in economic theory have not been validated by recent

---

*African countries as a reference point in this book pertain to the south of the Sahara.

evidence, which indicates that interest-rate differences have not
played a direct and major role in the growth of DFI in African countries. Rather, their role has been indirect through the effects of the
comparative abundance of labor and its low cost relative to capital.
Furthermore, the prohibition and the control of imports into some
of these countries and the relatively dependable access to sources of
mineral resources and vital raw materials that are abundant and ex-
ploitable by local low-cost labor are some of the more dominant and
direct factors explaining the recent growth of DFI in African countries. However, as a result of widespread pessimism about the postindependence political climate in Africa, there was a temporary decline in the level of total DFI in the early 1960s. While the past
steady increase has been resumed, there has also been increasing
concern by independent African states over the pattern of ownership,
control, and cost of foreign investment, measured, for example, by
inequitable compensation for the drain of material resources, profit
repatriation and other forms of remittances by foreign subsidiaries
and affiliates to their parent companies.

A major factor in the growth of capital formation since the
1960s is the reinvestment of profits. The foreign control of the private sector in African countries always has been preserved under
the guise of majority foreign capital inflow, but available evidence
for a number of firms, primarily in the extractive industries, suggests that their profits available for contribution to capital formation from within, specifically from the earnings of subsidiaries or
affiliates, now surpass the new, foreign capital inflow from the home
countries of the parent companies. In the case of foreign affiliates
of U.S. companies, internal U.S. sources as a percentage of their
total financing of expenditures on plant and equipment began to show
a definite decline in the 1960s. (See Table 1.1.) The broader evidence seems to indicate excess of earnings over foreign capital inflow to host countries. In the case of the United States, the decline
in capital outflow was accounted for, in part, by government restrictions on capital outflows since 1968. There is some evidence that
the reverse of this new trend holds for European companies. In the
latter case, there is little evidence that profits are all being rein-
vested. Instead, there are indications of profit repatriation; however, some appear to have been rerouted as new capital inflow to the
host countries.

It is customary for international companies to repatriate the
profits of their affiliates net of dividends. Foreign companies,
however, are finding it increasingly difficult to justify this practice,
more so in the face of the economic and political realities in the host
countries, specifically the increasing use of investment codes to
demand significant reinvestment of profits.

TABLE 1.1

Plant and Equipment Expenditures by the Foreign Affiliates of U.S. (Multinational) Companies and the Financing of Direct Investment, 1966-72

(billions of dollars)

Year	Total Plant and Equipment Expenditures	Earnings Reinvested	Internal Including Undistributed Profits and Depreciation	Summary of Sources of Funds for a Sample of Majority-Owned Foreign Affiliates of U.S. MNCs		
				Other Non-United States	United States	Total
1966	8.6	1.7	3.5	2.4	1.7	7.6
1967	9.3	1.6	3.9	1.9	1.0	6.8
1968	9.4	2.2	4.7	3.2	.5	8.4
1969	10.8	2.6	4.8	4.2	.9	9.9
1970	14.1	2.9	5.1	4.8	1.7	11.6
1971	16.3	3.2	6.3	5.2	2.2	13.7
1972	16.7	4.5	7.1	4.7	1.5	13.3

Sources: Survey of Current Business (Washington, D.C.: U.S. Government Printing Office, September 1971), table 1, p. 28; (October 1971), table 3, p. 28; (July 1975) table 1A, p. 30.

Since the mid-1960s, local economic control has become a national objective, and the flow of DFI has come under the jurisdiction of varying forms of "investment codes." These codes are intended to monitor and direct the inflow of new DFI into projects in priority sectors, supervise the gradual reduction of existing DFI, and promote indigenous equity participation in the foreign firms affected by the codes, but more importantly in all phases of economic activity. Since priority sectors, as outlined in national development plans of most African countries, ordinarily include those sectors in which the promotion of import substitution and export expansion are foremost, the increasing regulation of DFI has serious implications for the inflow of foreign technological innovations that could be vital to the development of those sectors. Before examining these implications, first, the pattern and the nature of DFI, including the conditions that manifestly impeded its optimum technological contributions in African countries, will be discussed.

DIRECT FOREIGN INVESTMENT

Ordinarily, exports represent the initial response to foreign demand. The next two alternatives usually are licensing and DFI. In African countries, the prerequisites--for example, institutional, manpower--for the broad application of licensing arrangements are either very limited or nonexistent. Consequently, once standardized export products are satisfactorily market tested, and assuming efficient national or regional market size exists, DFI is the more likely phase of entry than licensing arrangements.

The greatest surge in the flow of DFI to DCs has occurred since World War II and the major component of the flow to African countries has been into the extractive industries and "manufacturing," or assembly-type activities.* See Figure 1.1 for its distribution by sector and country of origin. In terms of the functional categories of DFI, the dominant component is the extractive industries. Services, including transportation, banking, advertising, and insurance, and manufacturing have developed in an ancillary role to the extractive industries. Although banking and insurance have

---

*In large measure, the agricultural sector in African countries remains traditional, and the availability of land and labor proved for centuries to be the basic input requirements until the development of farm machinery and the "green revolution." In spite of these developments, capital investment in agriculture has been generally low in comparison to the extractive and manufacturing industries.

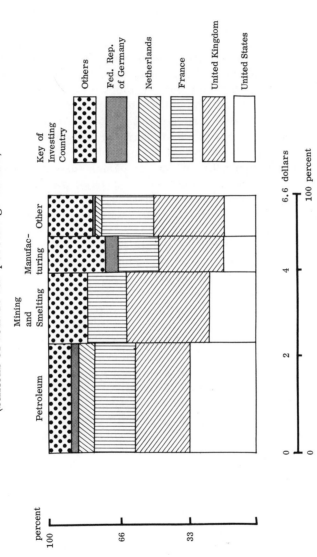

FIGURE 1.1

Distribution of Foreign Direct Investment in Africa by
Sector and Country of Origin at the End of 1967
(billions of dollars and percentage shares)

Source: United Nations, Department of Economic and Social Affairs, Multinational Corporations in World
Development (New York: United Nations, 1973), p. 18.

existed along with the trading companies in the early period of the colonial era, they have grown considerably recently in response to the growth of activities in the extractive and manufacturing industries and the concomitant economic and social changes. Some of the quasi DFI that took place in these sectors went to the distribution facilities of foreign trading companies. The extractive industries, including petroleum and other related activities, account for about 70 percent of total DFI. The growth of DFI in manufacturing since the 1960s has been spurred by numerous factors, some of which have been identified above. The primary ones are: the need of the foreign investors to maintain and to protect sources of raw materials; the relatively low wage rates that still typify many DCs; and increasing barriers to trade--for example, import controls and high tariffs --the changing pattern of export markets in the DCs, and the need of foreign companies to retain these markets and explore new ones.

Notable among the foreign companies responsible for the growth in DFI inflow, in addition to the major oil companies, are Bata Shoe Company, CFAO, Coca-Cola, Ford Motor Company, General Electric, General Motors, IBM, Imperial Chemical Industries, Lonrho, National Cash Register, Nestle, Phillips, Singer Sewing Machine Company, and Unilever. Unilever is the single largest source of DFI in Africa and epitomizes multinational vertical as well as horizontal integration.

Since the late 1960s, major U.S. firms like Bethlehem Steel, Alcoa, and Kaiser Aluminum, and the Japanese Nippon Mining Company have joined the rapidly increasing list of foreign direct investors in Africa primarily in the extractive industries. According to 1967 data, Great Britain accounted for the largest share, 35 percent, of total DFI in African countries, followed by France, 26 percent, and the United States, 15 percent.[1]

Historically, Italy, the Netherlands, and Belgium have also been important sources with 4, 5, and 7 percent shares respectively. (See Table 1.2.) Belgium's vast economic interest in the Congo, now Zaire, was interrupted temporarily after the colony's independence in 1960. While British, French, and Belgian DFI were concentrated in their former colonies, Italy's was relatively diffused. Until 1972, when Niger, Togo, Upper Volta, Chad, Cameroon, and the few other countries in the Franc zone began to exercise direct control over their respective economies, France had exercised considerable economic and political dominance over these countries, despite the attainment of political independence by most of them. A common currency in the Franc zone and the close identification of French interest with economic activity in each of the countries in the zone served to promote the growth of French DFI. U.S. direct foreign investment on any significant scale is relatively new in Africa.

TABLE 1.2

Stock of Foreign Direct Investment in Africa, from Development Assistance
Committee Countries, by Sector and Country of Origin, End 1967
(millions of dollars and percentage)

Sector	DAC Total (millions of dollars)	Sectoral Percentage of Total	All Countries	Distribution by Country of Origin (percentage)								
				United States	United Kingdom	France	Netherlands	Canada	Federal Republic of Germany	Japan	Italy	Other*
Petroleum	2,597.6	39.4	100.0	32.8	20.2	27.3	10.5	--	1.6	--	6.5	1.1
Production	1,947.7	29.6	100.0	33.7	20.2	29.5	10.7	--	2.1	--	3.4	0.4
Mining and smelting	1,279.8	19.4	100.0	20.9	36.1	22.2	--	0.8	3.5	0.0	1.8	14.7
Agriculture	496.8	7.5	100.0	10.2	18.1	51.5	4.1	--	1.1	0.2	0.6	14.2
Manufacturing	1,236.4	18.8	100.0	9.7	31.9	21.8	2.1	3.8	3.3	1.0	3.4	23.0
Trade	398.2	6.0	100.0	10.7	56.6	17.7	0.8	--	1.1	--	1.0	12.1
Public utilities	66.3	1.0	100.0	--	--	77.4	--	--	--	--	--	22.6
Transport	221.8	3.4	100.0	0.5	45.9	17.6	0.5	--	--	--	1.8	33.7
Banking	140.2	2.1	100.0	8.9	51.4	25.0	0.4	--	0.4	--	2.5	11.4
Tourism	43.7	0.7	100.0	57.2	11.4	27.5	1.1	--	2.8	--	--	--
Other	110.3	1.7	100.0	0.9	90.5	4.5	--	--	--	--	0.9	3.2
Total	6,591.1	100.0	100.0	20.8	30.0	26.3	4.9	0.9	2.1	0.2	3.8	11.0

*Includes Australia, Austria, Belgium, Denmark, Norway, Portugal, Sweden, and Switzerland.

Source: United Nations, Department of Economic and Social Affairs, Multinational Corporations in World Development (New York: United Nations, 1973), p. 180.

Of the $105 billion U.S. DFI worldwide in 1973, the most recent an-
nual data available, about 60 percent is located in Central and South
America, and only 2 percent or about $2.3 billion in Africa.

The nature and magnitude of the direct contributions accruing
to DCs from the participation of DFI in their respective economies
have become more controversial in the newly unfolding international
economic order. In defense of business practices by foreign firms
in these countries, and in delineating their roles and contributions
to host countries, the inflow of technological factors--abstract technology and the attendant know-how for its direct practical application
has often been cited as a vital part of an integrated flow of direct
foreign investment. In response to this claim, it has been argued
that DFI as an economic interest perpetuates dominance and dependence in DCs, and its attendant technological factors are instruments
of economic exploitation. In refuting this assessment, dominance
and dependence among unequal states are claimed to be better explained by power politics than by economic interests.[2]

Before the 1950s, studies seeking to ascertain the net effect
of the contributions by DFI, after accounting for the costs of their
existence and operations in the DCs as measured by wide-ranging
repatriation of profits, royalties, and other payments abroad, were
not conclusive for a number of reasons.

Virtually all African countries were and still are highly dependent on the production of primary agricultural commodities and,
since the 1950s, the agricultural sector has attracted much less
DFI than the extractive and "manufacturing" industries. (One exception is the participation of DFI in large-scale agricultural projects such as oil-palm plantations and rubber estates, but this is
also on the decline.)

This may be explained partly by colonial, and in some cases,
postindependence national policies that have tended to exclude DFI
from certain areas of agriculture, and partly a reflection of the low
rate of return in the exploitation of low-cost labor in agriculture
relative to the two other sectors. Although there is empirical evidence of a general nature about the flow of DFI into extractive industries, and a few consumer production activities, including distribu-
tive trades in African export markets, these have yet to be thoroughly
documented and their full impact assessed.[3]

So far, efforts from studies aimed at providing empirical evidence on the long-term contributions by DFI have focused primarily
on oil-producing DCs. The partial evidence available relied on
misleading indicators like payments to host countries in the form
of royalties and wages, taxes to the local government, and the
volume of sales by the foreign firms.[4] Beyond the apparent and
immediate impact of contributions in factor payments, the proper

assessment of the impact of DFI on long-term technological development of these countries, is likely to provide a more representative and complete measure to compare with the social opportunity cost of the total economic activity generated by DFI. Unfortunately, in addition to the problem of measurement, the peculiar vertically integrated structure of the firms in the oil industry, in their national and international corporate pursuits, has not been conducive to compiling any conclusive empirical evidence, thus leaving the controversy unsettled. The evidence provided in Chapter 7 on the contributions by foreign oil companies to the Nigerian oil industry is limited in scope, but hopefully represents a new departure. Realistically, it is doubtful that the controversy would ever be settled completely because of its inherent ideological and conceptual problems. Nonetheless, the technological dimension of DFI as it relates to African countries is examined variously in this volume.

THE FLOW OF TECHNOLOGY AS AN INTEGRAL PART OF DIRECT FOREIGN INVESTMENT

A widely recognized corollary of direct foreign investment flows to developing countries is the flow of technology, new production techniques and processes, as well as managerial and organizational know-how. Technology in this broad sense denotes the sum of knowledge--scientific, empirical, artistic--or, in general, experience and skills necessary for manufacturing a product or products and for establishing an enterprise for this purpose.[5] Technology flows through embodiment of knowledge in capital equipment and intermediate goods, techniques or processes, skilled manpower, and through the flow of proprietary and nonproprietary information. Virtually all of the technology flows from the highly industrialized countries to the DCs are proprietary in that when technology is made "available," it is customarily paid for by the recipient. Technology has been and continues to be a vital input in the economic development process. Unfortunately, in addition to the rapid socioeconomic changes it constantly generates, its political abuse has rendered it more and more susceptible to suboptimal use as well as misuse.

The importance of technology as an integral part of DFI and its capability of contributing to long-range economic transformation and growth in real income for each person have yet to be thoroughly examined and understood in the formulation of policy objectives for technological development in African countries. Furthermore, the successes and failures encountered in the flow of technology as a constituent part of DFI have yet to be fully documented. In essence, the questions--"When is a technology 'transfer' successful?" and

"How does one measure this success?"--are rarely posed. The
studies contained in this volume are intended primarily to fill a modest part of this void, and answer the first of these two questions.
The second question is a subject of current research by this author.

     The gap in technical knowledge and economic progress between
Africa on the one hand and Western Europe and the United States on
the other, though of relatively recent origin, has been widening considerably. Technological development in west Africa was disrupted
to a significant extent by two and one-half centuries of commerce in
slaves and by the eventual reorientation of the economy to colonial
requirements. Precolonial "manufacturing" activities in west Africa
had included cloth weaving, metal working, ceramics, construction,
and food processing.[6] Unlike the relationship among the advanced
industrialized countries in the flow of technology through the exchanges of technical, innovative information, the relationship between ACs on the one hand, and DCs on the other, consist of a uni-
directional flow of technology from the former to the latter group of
countries. The need to minimize the existing gap in technological
knowledge between ACs and the DCs (through the importation of purposeful technology by the latter) has been articulated and studied in
some depth from the former. However, no uniform opinion exists on
the inevitability of modern technology flows to DCs.

     It has been suggested by one school of thought (represented by
H. W. Singer) that the "transfer" of foreign technology to DCs is
essential and clearly beneficial to their economic progress. Others
contend that its costs to the importing DCs far outweigh its benefits
in view of considerable accrual of economic rent to its "suppliers"
that is made possible by the monopolistic advantage they command,
for example, through patents. The apparent inability of many DCs to
absorb efficiently the specialized technology developed specifically to
meet the input requirements in highly industrialized nations has led
to two other approaches.

     One of these suggests the development of "appropriate" or
"intermediate" technology, or what E. F. Schumacher, its chief
proponent, has recently described as "technology with a human
face."[7] The "virtues" that modern technology lacks, but that E. F.
Schumacher's technology would possess, include "the capability of
being self-balancing, self-adjusting, and self-cleansing." Its character is organic rather than a foreign body with unremitting suscep-
tibility to rejection. Schumacher's proposed approach represents a
significant departure from an orthodox approach to the analysis of
the role of science and technology in the promotion of economic
progress.[8] The second of the two approaches seeks to promote
technological self-reliance among DCs. It advocates a shift in emphasis from foreign to local development of technology for the fullest

possible exploitation of available domestic resources. The requirements of this approach invariably entail massive initial outlays for
the importation of intermediate inputs. [9] This approach also lacks
any promise of a solution to immediate economic problems in the
DCs, and it has the obvious tendency to create another form of dependence that, if not transitory, becomes totally inconsistent with
the ultimate objective of the approach itself.

Granted that these two approaches have merits as perceived by
their proponents, it is about a decade now since they were first advanced, and no significant change has occurred in the character and
manner of the technology flows to DCs, and there has been no apparent major progress in indigenous technology development.

Based on various empirical evidence, only about 5 percent of
the technology currently being used by DCs in Latin America and
Asia for productive activities in industry is wholly indigenous. Virtually all of the other 95 percent is accounted for by imported foreign
technology primarily from private firms. This distribution has not
changed considerably in the last two decades. In the case of African
countries, almost all of the modern technology in use in industry,
manufacturing and extractive, are imported. The heavy dependence
on foreign technology by DCs in general has been considered re-
mediable without a complete and sudden shift to the development of
new and wholly indigenous technology. Clearly, the development of
technology anew is expensive, extremely time consuming, and risky.
These characteristics of the economics of technological innovation
define some of the limits to self-reliance and underscores the inevitability of varying degrees of dependence on foreign technological
knowledge, even for the development of indigenous technology.
These requirements also have to be reconciled with another com-
pelling consideration, namely, the sociopolitical necessity of developing enough indigenous productive capabilities for local control of
the national economy in the shortest time possible. It is in light of
these assessments that the role of the foreign technology associated
with DFI is examined to ascertain the forms of its contributions to
technological development in African countries.

The supply of technology from firms in industrial countries and
its importation by DCs may be considered as two different processes
involving active as well as passive roles on the part of the suppliers
and the recipients. The "transfer" of technology through DFI is, by
its nature, highly restrictive and expensive. Although it provides
the prospective recipient DCs with the opportunity to take advantage
of the fruits of heavy expenditures by foreign firms in the trial and
error of research and development, or R and D, and in subsequent
technological innovations, the private and social benefits are generally subject to major limitations. The transfer is usually manifest

in the form of passive "borrowing" or temporary acquisition because of limited choices imposed explicitly and implicitly by past colonial ties, the general lack of necessary information for consistent rational choices, and the "channel capacity" in the individual DCs. Channel capacity is the limitation that the user of a technology faces in his or her ability to handle and effectively use pertinent technological information. In most African countries, this capacity is considerably more limited than it is in other DCs in Asia and Latin America because of the generally passive posture of the scientific community in the application of science and technology to solve society's problems.

Therefore, technology transfer tends to be superficial. The acquisition process is constrained by the limited ability and capacity to discriminate and acquire what is desired based on limited available information. The serious dilemma emanating from this condition is that in exceptional cases, where opportunity exists for complete acquisition, the indigenous scientific and technological manpower necessary for effecting appropriate choices for absorption and diffusion, is generally in short supply.

Thus far, an important conceptual problem has been ignored and technology transfer has been treated cautiously. Next, some of these problems are examined from the viewpoint of the DCs. In order for the concept of technology transfer to be meaningful, attempts must be made to answer the first of the two questions raised before. Namely, when is the transfer of a technology successful? In such attempts, a distinction ought to be made between the "transfer" and the acquisition of technology. When the suppliers and the recipients of a foreign technology play active roles in the flow process, technology transfer and acquisition become the single purposeful process of evaluating, purchasing, reevaluating and adapting (when necessary) technical information from one institutional setting (that is, the suppliers) for similar or different end use in another institutional setting (that is, the recipients). Under this condition, the technology flow is likely to be more responsive to the most pressing needs as well as conform to the prevailing conditions in the recipient country. The flow, therefore, extends beyond the mere dissemination of technical information and may be independent of foreign equity participation. How positively correlated is the extent of foreign equity participation with the effectiveness of the flow of technology to DCs? This is a legitimate empirical question, but it is also a difficult one to resolve without reliable empirical data that thus far have been hard to obtain. However, it is generally acknowledged that the technological knowledge associated with the inflow of DFI into Africa, the bulk of which went into extractive industries is generally proprietary and has tended to be isolated from the rest of

the economy. The flow of technology as a constituent part of DFI
through subsidiaries or joint ventures is typically a "composite" or
"packaged" flow consisting of physical capital equipment/machines,
intermediate inputs, and technical knowledge of foreign "experts."
The last component of this "composite" flow has generally lacked
the complete freedom necessary to perform the types of tutorial
roles that could facilitate "learning" and subsequent takeover of
major technical functions by indigenous manpower. A "composite"
transfer may or may not be desirable, depending on the prevailing
circumstances. In most African countries, conditions are such that
"composite" technology transfer has tended to perpetuate major technological and organizational dependence. Associated with "composite" transfers are a high rate of technological innovations but
an almost complete dependence on the organizational and marketing
facilities of the parent company and, consequently, on expatriate
personnel. Furthermore, the importation of intermediate inputs
from the supplier of the technology is usually a prominent feature in
the arrangement. In sum, the typical flow either does not use, or
makes inefficient use of, local substitute resources. This represents
a major imposition on the attempts to cope with domestic unemployment problems in most DCs.

The basic asymmetry in the conduct of business operations by
foreign firms and the pursuit of national development objectives by
host governments is evident in conflicts about policies on the appro-
priateness of transferred technology, consideration for local conditions in terms of manpower requirements and development, long-
term technological development, and ecological equilibrium. The
complex channel through which technology flows and the differences
in the objectives of its suppliers and recipients has also meant that
what the technological component of DFI is capable of contributing
and what it is contributing are two different things.

This discrepancy between the actual and the potential contributions may be explained again by the channel capacity in the DCs and
more importantly, by the restrictions against the requisitions of
technological know-how. To justify their presence and securely retain controls over their operations in DCs, foreign companies, particularly those manufacturing sophisticated products or possessing
sophisticated processes, explicitly treat the flow of technology as an
integral part of DFI and effectively maintain secrecy of their technological edge by ensuring that they retain full or at least controlling
equity in their subsidiary or the local enterprise involved. Although
the scientific knowledge, the "know-why," to be applied is generally
in the public domain, the technology, the "know-how," necessary for
its application and the capital requirements for research and development is neither readily available to African indigenous investors,

public and private, nor fully accessible for nonrestrictive nonproprietary acquisition.

THE "SALES" AND "PURCHASES" OF
TECHNOLOGICAL KNOW-HOW

It is apposite to conclude from the foregoing that the flow of
technology from industrial countries to DCs rarely culminates in outright "sales" and "purchases" of know-how. Rather, the technology
is "rented" under highly restricted conditions analogous to those
faced by tenants in rental accommodation. Typically, under these
conditions, changes in the physical structure of imported machines
in which a significant part of the technology flows are ordinarily embodied may not be possible through the careful commercial protection of the technological knowledge or by outright legal measures to
prohibit such changes. These conditions also preclude the feeling of
ownership and the direct responsibility associated with it. Consequently, a passive posture is often taken by the technology recipients,
and all repairs to the imported machines become the responsibility
of the renter-owner. All the critical issues surrounding the supply
of technology to DCs must be examined within the implications of
this analogy for significant conclusions on the likelihood of nurturing
imported technology.

Until the late 1960s, the untrammeled operations of foreign investors have tended to foster "supply-push" rather than "demand-
pull" conditions in the flows of technology. An explicit delineation
of the differences in the flows of technology among ACs, and from
these countries to DCs is generally absent from the body of literature on this subject. Conceptually, supply-push conditions prevail
when the transfer of proprietary and nonproprietary technology is
solely a supplier decision and the structural conditions in the recipient's environment are of no primary consideration in the transfer.
Demand-pull forces are at work when the supply of technology is in
response to the needs of the recipient and the conditions in the recipient's environment are the primary consideration in the importation and acquisition. This distinction between supply-push and
demand-pull technology flows suggests the foregoing implied parallel
between the transfer and importation of technology respectively.
The apparent disaccordance between the supply-push and the demand-
pull decisions in the flow of technology also has a historical parallel
in the flow of bilateral, governmental technical assistance from the
ACs to DCs. Historically, assistance in the latter case has been
provided under bilateral arrangements based primarily on the perceptions of the donor country and in its best political and economic
interests.

Being a "sellers' market" and supply-push in character, the flow of technology often entails the borrowing of the latest technology. Apart from its relatively high price, the advantage of economies of scale typically associated with a latest technology is often sacrificed because of the small extent of market and restricted access of the recipients to export markets. One further distortion, typically compounded by public policy on incentives for foreign investment expansion, is intensity in the use of scarce capital resources due to their artificially low price, while relatively abundant labor resources are underemployed. Consequently, the narrow options in the choice of technology open to the recipient countries have meant serious financial and technological impositions, which are also in large measure attributable to transfer pricing, adverse terms of trade, and the supply-push character of technology transfer.

Since the early 1960s, the effects of foreign technology acquisition on the balance of payments of numerous DCs in Latin America and Asia have been examined. In Latin America, for example, payments for licenses on patents, consultants and engineering services have been estimated to be generally low, usually not exceeding 5 percent of total imports of goods and services and the contract terms rarely are found to exceed 10 years.[10] This estimate and similar ones on DCs in Southeast Asia are misleading. Imported machinery and equipment typically presumed to embody technological know-how have been found in other studies to be overpriced, and thus contribute to the persistent balance-of-payments deficits in a large number of DCs.[11] In considering technology as an integral part of DFI, it is clearly erroneous to treat it as nonproprietary and free. Subsidiaries of foreign companies in practice make lump-sum payments for specific technological know-how and also make regular remittances to the parent company for research and development costs.

ALTERNATIVES TO DFI

In the interests of minimum complexity, the transfer of foreign technology through DFI has been examined above without any distinction made between the manner of the flows of DFI itself and the nature or process of the associated technology flows. But, major differences clearly exist between the two, and these are better outlined in an examination of some possible and potentially more effective alternatives to DFI in the supply and acquisition of foreign technology by DCs.

These differences are best illustrated by the following questions: What is the percentage share of DFI and the extent of control by the parent company in its subsidiary or affiliate? Is the investment an actual parent company to subsidiary flow or a rerouted flow

through another foreign firm? Is the technology associated with the
inflow of the DFI proprietary to the parent company, or has it been
obtained for its use and its subsidiaries' through licensing or management contract? The intent of these questions is not to provide
answers to them, because that can best be done only on a case-by-
case basis. Rather, the object is to present some of the complexities
involved in the transfer of technology through DFI and why it may be
relatively ineffective as currently practiced, compared to the alternatives examined below. In general, foreign investors, including
multinational companies, prefer DFI to outright licensing of technology to indigenous enterprises in the DCs because of the opportunity
for new markets that it provides. More importantly, DFI makes it
possible to maintain secrecy and control over the transferred technology and presumably over product quality and marketing.

To overcome the typical restrictions and some of the other
disadvantages associated with DFI-related technology flows, changes
in public policies affecting foreign investment are aiming more and
more toward the diversification of the inflow of DFI as well as the
sources of technology. Such policy changes invariably have to be
accompanied by rapid development of indigenous entrepreneurship if
a reduction in the domination of DFI is also a policy objective, as is
commonly the case. Already in Ghana, Nigeria, Tanzania, and a
few other African countries, DFI is disallowed from competing
against indigenous entrepreneurs in prespecified businesses and also
restricted from participating below certain minimum registered capital. Although there is no conclusive study on the social costs and
benefits of DFI, there is a general conviction among policy makers
and indigenous entrepreneurs in African countries that the social
costs far outweigh the social benefits. Of course, the measurement
problems involved in the cost-benefit analysis needed to verify this
opinion, are monumental.

As public policy toward foreign investment inflow broadens in
African countries, and the number of subsidiaries or enterprises in
which overseas parent companies or foreign investors have controlling ownership begins to show a steady decline, in response to investment codes' guidelines, as it is already the case in Ghana, Nigeria,
and Tanzania, technology flows through DFI should be expected to
assume even less importance.

Apart from DFI, another dominant mode of effecting technology
flows, particularly among ACs, is licensing. Foreign companies
periodically use licensing arrangements to effect the transfer of
technology to their subsidiaries in DCs, but not to the same extent as
the practice among ACs. The international licensing of industrial
property rights is by far the most formalized and commonly used
mechanism in enterprise-to-enterprise technology transfer. In

terms of its applicability, this mode has its own drawbacks and they vary depending on the state of industrial development in the technology importing country and the extent and sophistication of indigenous entrepreneurial capability it has in existence. An important precondition for effective transfer of technology through licensing agreements is a suitable scientific and technological infrastructure for the exploitation of the imported know-how.

The parties to a license agreement usually consist of the licensor and the licensee. According to the World Intellectual Property Organization, a license agreement is the contract between a licensor --the owner of an exclusive right granted under the law pertaining to the exploitation of a technical invention, for example, a patent--and a licensee on the granting of a license. The license is the consent given by the licensor to the licensee as to the use of a technical know-how for which a fee is paid in return.

Depending on whether there is equity participation by the licensors, the safeguards that licensing agreements provide the licensors include the opportunity to

- convert idle patents and trademarks into cash assets--lump-sum payments, periodic cash royalty income, sales of prototypes as well as related components and spare parts--without additional capital expenditure
- eliminate need for time-consuming and costly plant construction abroad
- increase capital inflow with minimum outlay
- test foreign markets at a minimum of expense
- reach and retain access to markets that might otherwise be lost through import restrictions
- penetrate foreign markets obstructed by, for example, high duties and "local sales peculiarities."

The manner in which licensors optimize the exploitation of a technology is not limited solely to those developed to support and protect the competitive edge of their companies. Technologies generated by internal research sometimes result in a spin-off that is of no direct use to the innovating company, and therefore find it necessary to seek out "purchasers."

Theoretically, though not necessarily in actual practice, certain advantages accrue to the licensee in the DCs. These include the opportunity to

- obtain costly, advanced and completely developed and tested technology, and know-how, of others with a minimum of investment and in a short length of time

- eliminate the need for extensive research otherwise necessary for improvement and innovation in existing industry
- obtain technology cheaper than developing their own technology
- obtain management and marketing assistance
- increase export potential and realize favorable balance of payments, and
- expand employment.

In practice, the disadvantages to the recipient country may be in the form of limited freedom in the use of the licensed technology and/or inordinate and fictitious costs. The following are some examples:

- downpayments may not be recoverable when it is found that the licensed technology is not adequate or is not suitable for licensee's market, and it may be "obsolete" or found to be too expensive at an advanced stage of the agreement
- accepting a license from one licensor often forestalls access to other licensors, and the licensee may have chosen the wrong technology or the wrong licensor
- licensee may be harmed by unlicensed competition if licensor's patent protection is weak
- in general, the type of relations between parent companies and subsidiaries may mean that the latter are not always free to buy appropriate, required, technology from other sources or search for alternative technologies. Under these circumstances, the prices paid for transferred technology are dictated by the parent company and, based on the "letter" of most licensing agreements, there may be little or no freedom of choice as to export markets.
- indigenous firms, so far, lack the resources to enable them to obtain information about possible alternative sources and arrangements concerning the supplies of the technology they require; such information would have enabled them to buy in a more competitive market. A similar situation exists with regard to government enterprises that are further handicapped by the fact that, for historical reasons, the officials who assume a leading role in decisions about technology "purchases" are often not knowledgeable about technological questions.

Notwithstanding these disadvantages, licensing warrants closer examination and consideration in operationalizing specific alternatives to technology transfer through DFI. Three nonrestrictive, or relatively less restrictive, but viable licensing alternatives to DFI are:

- strictly nonequity participating private enterprise to private enterprise transfer and acquisition of proprietary technology for a fee

- private enterprise to government enterprise transfer and acquisition for a fee
- direct government-to-government transfer of nonproprietary technology.

The last alternative, unlike bilateral technical assistance, primarily entails the transfer of technology on a nonexclusive royalty-free basis but applicable mostly to technological know-how on which patent rights have expired but are still appropriate, and suitable for the needs of DCs.

The first of these alternatives represents a major departure from DFI as a mode of transferring technology in that it does not provide for the maintenance of the type of complete secrecy that, in addition to affording the owners of the technological innovations with a "reasonable" return on their risk capital for R and D, also result in the accrual of economic rent, usually indefinitely. Although the first alternative does exist on a limited basis, the effectiveness of the third is likely to provide the necessary impetus for the broader practice of the first by potential suppliers of technology.

The shift from foreign to wholly domestic or majority equity participation and control that has been taking place in certain industries primarily in Nigeria, Ghana, Tanzania, and to some extent in Zaire, has come about not as a result of increases in private domestic investment but from rapid expansion in domestic public investment. The second of the three alternatives above has been the vehicle through which public enterprises avail themselves of foreign technological know-how. This approach offers opportunity for the servicing and implementation of well-defined projects, in terms of properly coordinated national development objectives, through the acquisition of technology by public enterprises, which generally are in a better bargaining position than private domestic enterprises. One example is the Nigerian National Oil Corporation, a public enterprise, and its relatively strong bargaining position in relation to the foreign oil companies in Nigeria. This subject is examined in detail in Chapter 7. In practice, arrangements between private foreign firms and government enterprises as illustrated by the second listed alternative have increased considerably in recent years, primarily as a matter of necessity. Until ample private indigenous entrepreneurial capability has been developed to fill the void being created by the decline in DFI and foreign entrepreneurships, the second alternative should be expected to increase even more in importance.

What then are the implications of the increasing regulation of DFI for the acquisition of appropriate technology? In the case of the most sweeping changes in foreign investment policies in African countries that began in Nigeria in 1973, there has been no conclusive

evidence demonstrating any major adverse effect. Contary to the
adverse initial reactions to the 1973 Indigenization Decree in Nigeria,
its mandated acquisition of majority control in a number of foreign
businesses by Nigerians, and its apparent maximal implementation,
has not been completely inimical to the inflow of DFI. Similar and
much more restricted investment policies are in existence in other
DCs in Asia and South America and experiences differ. The notable
ones are India, Mexico, and the Andean Group: Bolivia, Chile,
Colombia, Ecuador, Peru, and Venezuela.

The last group of countries, in practice, ties the inflow of
DFI to the inflow of appropriate foreign technology on terms acceptable to them. These major shifts in host nations' investment policies have given a different perspective to the nature of joint ventures
between foreign firms and government enterprises in many of these
countries. (A further discussion of these shifts is contained in
Chapter 3.) As DFI becomes less and less welcomed, except on the
mutual terms of the recipient countries and the foreign investors, a
necessary condition for the continuing inflow of technological know-
how is the development of a new set of partnership arrangements.
Three examples of such potentially viable arrangements are identified above. These three alternatives will be examined variously in
Chapters 3 and 4. First, in Chapter 2, Mira Wilkins examines the
broad historical background to the changes in the role of DFI and
the transfer of technological innovations especially before 1950.

NOTES

1. See Organization for Economic Cooperation and Development, Stock of Private Direct Investments by DAC Countries in Developing Countries end 1967 (Paris: 1972).

2. P. M. Sweezy and H. Magdoff, "Notes on the Multinational
Corporation," Month Review 5 (October 1969): 1-7 and (November
1969): 1-13. See also Benjamin T. Cohen, The Question of Im-
perialism: The Political Economy of Dominance and Dependence
(New York: Basic Books, Inc., 1973).

3. T. Bauer and Basil S. Yarrey, The Economics of Underdeveloped Countries (Chicago: University of Chicago Press and
Cambridge University Press, 1957).

4. See Lester F. Pearson et al., Partners in Development
(New York: Praeger Publishers, 1969), pp. 103-04; and Herbert K.
May, "The Contributions of U.S. Private Investment to Latin Amer-
ica's Growth," a report prepared for the Council for Latin America,
Inc., January 1970.

5. United Nations Industrial Development Organization, Guidelines for the Acquisition of Foreign Technology in Developing Countries (New York: United Nations, 1973), p. 1.

6. For example, knowledge of iron smelting existed in west Africa at Nok (northern Nigeria) around 500 B.C. See A. G. Hopkins, An Economic History of West Africa (New York: Columbia University Press, 1973), chapter 2, pp. 27-50; and Walter Rodney, How Europe Underdeveloped Africa (London: Bogle-L'Ouverture Publications, 1972), chapters 4 and 6.

7. See H. W. Singer, "The Development Outlook for Poor Countries: Technology is the Key," Challenge: The Magazine of Economic Affairs 16, no. 2 (White Plains, N.Y.: May/June 1973): 42-47; and E. F. Schumacher, Small is Beautiful: Economics as if People Mattered (London: Blond and Briggs Ltd., 1973). Schumacher is the founder of the Intermediate Technology Development Group. ITDG has been engaged in the development of appropriate production techniques and machines based on the observed conditions and needs of DCs.

8. Ibid., chapter 5.

9. See Goodwin Matatu, "Development and the Human Environment," in Models of Development (London: SCM, 1973), pp. 19-28, and Peter E. Temu, "The Concept of Self-Sufficiency," mimeographed (Dar-es-Salaam: East African Universities Social Science Conference, December 1970), pp. 62-68.

10. R. Hal Mason and Francis G. Masson, "Balance of Payments Costs and Conditions of Technology Transfers to Latin America," Journal of International Business Studies (Spring 1974), p. 73.

11. See United Nations, Economic Commission for Latin America, "The Transfer of Technology and its Relation to Trade Policy and Export Promotion in Latin America," Economic Bulletin for Latin America.

PART

I

DIRECT FOREIGN INVESTMENT AND THE TRANSFER AND ACQUISITION OF TECHNOLOGICAL INNOVATIONS

2

MULTINATIONAL COMPANIES
AND THE DIFFUSION
OF TECHNOLOGY TO AFRICA:
A HISTORICAL PERSPECTIVE
Mira Wilkins

Historically, there have been numerous and various multinational companies operating in Africa. They came in waves: first, the trading companies; second, the banks and insurance firms; third, the mining and agricultural ventures; fourth, the enterprises that made infrastructure-type investments; fifth, the petroleum companies; and finally, sixth, the manufacturers. The waves overlapped with one another. Each brought new technologies to Africa.

THE TRADING COMPANIES

Traders introduced previously unknown agricultural products. In the sixteenth century, the Portuguese brought from Latin America into west Africa, maize, manioc, sweet potatoes, peanuts, papayas, cayenne pepper, tomatoes, and tobacco. The cultivation of these products spread through tropical Africa. The Portuguese carried cacao, cocoa, to Sao Tome from South America in 1822; cocoa became, a century later, the principal export of the Gold Coast, now Ghana. European traders brought wheat to southern Africa, where its cultivation developed in the nineteenth century. Africans and Europeans raised the new crops with different production methods, although, as we will see, there was some diffusion--indigenous adoption--of European processes. What is important is that the traders transferred the product technology, that is, the crop itself.[1]

The seventeenth-century trading companies, the Dutch West India Company, the Royal African Company, the Compagnie des Indes Occidentales, and the Compagnie Royale du Senegal, Cap Nord, et Cote d'Afrique among them, were all slave traders. They sold in Africa, European goods such as textiles, beads, iron bars, and guns, and stimulated taste for these items. The Royal African Company was reported to have encouraged the cultivation of indigo, a new crop for Africa, and the manufacture of potash. This company sought to develop a traffic in locally made cotton goods between Sierra Leone and the Gold Coast.[2] That these trading firms introduced new products, new trading techniques, and new methods of cultivation and manufacture, did not mean much in terms of effective technological diffusion. The European-manufactured goods sold in Africa were not copied and produced there, nor were the trading techniques imitated. Existing internal African trading networks proved durable.

Similarly, in the eighteenth and nineteenth centuries, European trading firms prompted new demands in Africa, but other than new crops, a significant technological contribution, the amount of technology actually absorbed within African societies proved limited. When, in the nineteenth century, "legitimate commerce" replaced the slave trade, when British machine-made cotton textiles and iron goods flooded African markets, new manufactured products became known, but the transferred merchandise once again did not result in import substitution--that is, domestic machine-made textile production or European-type iron manufacture. The conditions were not ripe for local factory production to emerge to meet local needs.

In time, trading companies became instruments for colonization. The Company of Merchants Trading to Africa, which in 1750 had obtained the "liabilities and responsibilities" of the Royal African Company, received British parliamentary permission to govern conquered settlements, and in 1765 the company's "sphere of influence," Senegambia, became Britain's first African colony.[3] Commerce seemed to require the maintenance of order--and sovereignty followed trade.

The late nineteenth century saw the chartered companies penetrate Africa.[4] Unlike the earlier trading firms, the chartered companies ventured beyond the coast into the interior; they had broader plans than mere commerce.

Among the companies, with their dates of charter cited, were: German Company for South-West Africa (1885); German East Africa Company (1885); Royal Niger Company (1886); Imperial British East Africa Company (1888); Comite d'Etudes du Haut-Congo (1888); British South Africa Company (1889); Mozambique Company (1891).

The chartered companies were generally granted by European heads of state specific rights in land, development of minerals, and also an administrative obligation. They obtained control over vast territories. They were authorized to trade within Africa and abroad, but they were far more than mere trading concerns. In this chapter, we will see them in their numerous functions. As corporate sovereigns, their technological contribution related to social organization. They transferred new forms of administration to Africa.

From the late nineteenth century on, trading companies, large and smaller ones, brought to Africa a sequence of products that proved more varied than yard goods or simple iron wares. Bicycles, cigarettes, soap, radios, in turn, created new values and stimulated new needs for money.

But more importantly, to pay for the imports, the new wants prompted the development of cash crops. The trading companies provided international markets for these crops. Primitive production methods often prevailed, with little transfer of European methods of cultivation to the native sector. The technological change lay in the move from subsistence farming to the cash crops.

In short, over the years, European traders--there were relatively few American traders involved--introduced technological innovations, some of which were absorbed, from peanuts to cocoa; some of which shattered traditional relationships, from new administration to cash crops; and some of which provided new dreams and hopes for the future, from bicycles to radios; but all of which failed to provoke effective imitation or significant transfer of new methods of production.

THE BANKS AND INSURANCE FIRMS

Of all the multinational companies, the banks probably had the least direct impact on the transfer of technology. Yet, indirectly, they also had significance. To the extent that trade and other business operations required banking facilities in their technological transfers, the banks played an important role. Likewise, if one defines technology broadly--as we would like to do--to include the structure of business organizations, the banks can be said to offer an example. The first multinational bank in Africa was the London and South African Bank, opened in Capetown in 1861.[5] Others followed.

So too, insurance firms moved into Africa. New York Life Insurance Company started to do business in Algeria in 1880 and in Cape Colony in 1883.[6] The company introduced innovative procedures. One has the impression, however, that it catered to Europeans instead of Africans and had little impact on changing the technological--broadly defined--base of African society. Its impact on tropical Africa appears to have been negligible.

The banks and insurance company followed a pattern that we will see often repeated; they entered Africa, through South Africa or north Africa. Later, they would add to their ventures and develop business in other parts of Africa.

It would be useful to have a study of multinational banking and insurance that would look specifically at their historical, technological impacts on less developed countries. To this author's knowledge, no such studies exist. Such research might take into account the technological role of the large European-based financial groups, for example, Societe Generale in Belgium, the Great German banks, and the Rothschilds in France and Britain. It would be very useful to have substantial additional inquiries that relate the activities of the financial institutions to technological transfer and diffusion.

MINING AND AGRICULTURAL VENTURES

Mining and agricultural enterprises had clear and substantial technological consequences in tropical Africa. These mining and agricultural activities were primarily by European men and capital and only to a small extent by their American counterparts.

Before World War II, the sequence in significant mining was first diamonds, then gold, and next copper. Of less importance were asbestos in Rhodesia, chrome in South Africa and Rhodesia, tin in Nigeria, and manganese in the Gold Coast. The post-World War II developments included iron ore, bauxite, and a range of specialized minerals.

The era of African mining concessions started with diamonds. From 1867, when the first diamond was discovered in South Africa, to the end of 1936, the value of diamond output in South Africa alone exceeded £320,000,000, an amount that surpassed the value of all minerals produced up to that time in the rest of Africa. [7]

By 1881 there were 71 private companies either involved in or floated for the purpose of South African diamond mining. Key among them were the De Beers Mining Company and Compagnie Francaise des Mines de Diamant du Cap de Bon Esperance. [8]

When we look, in general, at mining ventures, several types seem in evidence. The first type typically emerged out of an existing trading activity. The French diamond enterprise mentioned above fits into this category. It was started by a representative of a French firm of diamond buyers. So too, the De Beers Mining Company was organized by Alfred Beit, who represented a German diamond buyer. The diamond industry required little outside capital. It provided its own surpluses to finance its growth. Cape Colony residents were the early investors. [9]

In 1887, Cecil Rhodes acquired complete control of the De Beers Mining Company. Soon, he was amalgamating the major mining companies in the diamond industry, under the De Beers name. Diamond mining spread west and north. The De Beers Company developed mines in Southwest Africa, the Belgian Congo (now Zaire), Angola, the Gold Coast (now Ghana), and Sierra Leone. The De Beers firm, along with its affiliated enterprises, controlled the world's diamond industry. Separate diamond-mining companies sold through the De Beers network. [10]

A second type of mining operation can be placed under the heading of what John Stopford, an expert on international business, has called "expatriate investments."[11] "Expatriate investments" seem to represent an important aspect of British, and apparently other European, private foreign investments. Like the investments of today's multinational corporations, these carried management and technology, but unlike most direct foreign investments of today, they did not grow out of the foreign expansion of domestic manufacturing or mining companies.

Expatriate investments were often listed on London or other European exchanges, attracted portfolio investments, and are usually grouped into that category in the literature on overseas investments.[12] Yet, it is important that these investments were not simply capital transfers. The enterprises established did introduce foreign management and control. For this reason, the author believes that they should be included in discussions of direct foreign investment, that is, in discussions of multinational enterprise.[13] Like other multinational corporations, they were conduits of technology.

In a sense, the De Beers Mining Company, which raised some money in Europe, came to fit under the rubric of "expatriate investment." Yet, since it grew from European trading ventures, from European domestic operations, it would seem to belong in its own category. Obviously, the line is not distinct. Beginning in 1881, with seven gold mining and exploration companies, all registered in England, the Gold Coast, and, to a lesser extent, other locales on the west African coast, attracted 82 gold mining companies in 1900 and 220 new ones in 1901 alone.[14] These were all "expatriate"-type companies. Meanwhile, in the Rand, in South Africa, diamond merchants and entrepreneurs became interested in gold. The first South African gold mining company registered in Great Britain was the Gold Fields of South Africa, Limited, 1887. Cecil Rhodes was associated with this company.[15] From the late 1880s onward, such expatriate investments became common in Africa, whether it was the Rhodesia Chrome Mines Limited, formed in 1908, or Fanti Consolidated Ltd., formed in 1910 to mine in the Gold Coast.[16] Some of the expatriate investment involved a cluster of companies, some in mining and some in other sectors. Thus, the British financial trust, the Zambesia Exploring Company, had large interests in the Benguella Railway, the Tanganyika Concession, Union Miniere du Haut Katanga, and the Rhodesian Railways. The Tanganyika Concession in turn possessed substantial interests in the Katanga mines and worked in concert with the Zambesia Exploring Company and Union Miniere. A large venture was Ernest Oppenheimer's Anglo American Corporation of South Africa, which started in 1917 in gold mining and then sought diamonds, copper, coal, and other metals.[17] Its mining projects spread out from South Africa. It was far from confined to South African mining.

In the Belgian Congo, mining companies were frequently associated
with large Belgian financial groups. For example, by 1932, Societe
Generale--a giant financial group mentioned earlier in which the
Belgian government was a shareholder--controlled 41 companies in
the Congo, of which the largest number, 12, were in mining.[18] In
general, with the expatriate investments in mining, the corporate
shell would be related to the imperial, that is, colonial, power. The
funding was generally of the nationality of the imperial power, although there were a number of cases where other European, or
American capital participated. We need more evidence on how these
mixtures came about, their effect on management, which generally
seems to have been from the country of the imperial power, their effect on the technical staffing, and most importantly, for purposes of
this article, the impact of the mixed venture on technological transfer. Technology in mining was international. In mining, for any one
venture, it is not clear that there was a necessary coincidence between the sources of capital, management, technical personnel, and
technology of mining. Yet, all four were brought together in the
single mining venture.[19]

A special type of expatriate investment in mining lay in the
chartered company. A number of the chartered companies obtained
rights to develop mining areas. Economic historian D. K. Fieldhouse, for example, writes that the British South Africa Company
"began as a mining concession which Rhodes' agent Charles Rudd ex-
tracted from Lobeguela, King of the Matebele." Rhodes floated this
company on the London exchange. There was a London office, a
"glittering board of directors" designed to attract British money,
but the control was by Rhodes, who ran the company from South
Africa.[20] The company operated in what is now Rhodesia and Zambia.

Particular individuals' names run through the literature on the
British mining firms, Cecil Rhodes, Edmund Davis, Ernest Oppenheimer. Each was a director and shareholder in a large number of
mining companies, whether those stemming from trading companies,
"expatriate investments," or linked with chartered companies.

Another distinguishable type of mining venture was tied in with
already existing multinational mining activities. Here too the line is
muddy. Do we include the international mining operations--after the
initial investments--of Rhodes, Davis, and Oppenheimer? None of
their companies appears to have had any mining in its home country,
Great Britain. By contrast, there were international mining companies with headquarters in particular home countries that did have
home-based mining and those foreign businesses were involved in
horizontal integration. Thus, the Guggenheims, headquartered in
the United States and active there (and internationally) in mining and
smelting, made direct investments in the Belgian Congo (through

Societe Internationale Forestiere et Miniere du Congo) and in Angola (through Companhia de Angola). Their investments in the Congo dated from 1906. In both the Congo and Angola, the Guggenheims joined with European capital. They discovered and mined diamonds in the two countries.[21] American Metal Company, a U.S.-headquartered multinational--after World War I, developed interests in the late 1920s in copper mining in Northern Rhodesia, now Zambia.[22] The British-headquartered Anglo-Oriental Mining Corporation had tin mines in Cornwall, at home, and in Burma, Malaya, and Nigeria.[23] It is not clear whether there were marked differences in technological transfers when the company had a home-based mining operation and when it did not. Tentatively, this author would like to hypothesize that there was no difference, because of the international character of western mining technology. This would, however, make a fascinating research study. What would be interesting to see is whether the expatriate mining companies, with no home-based mining, were able to buy the technology and transfer it, or were able to spread the technology from one overseas venture to another, or whether they had difficulties in this sphere. What kind of difficulties did they have? Were companies that had home-based mining more successful in transferring technology, because of their own home-based knowledge? In many ways the questions are highly pertinent to today's problems, because they deal with how technology moves between and among different economic units. We need far more study of the relationships of ownership, in these mining operations, and technological transfer.

Another type of mining operation in Africa involved the vertical, or backward, integration of existing multinational manufacturing firms. Sometimes the particular manufacturer had experience in the type of mining and sometimes, for the particular manufacturer, the mining was a new business and the technology, like that of the expatriate mining operation, had to be newly acquired. How much mining in Africa involved backward integration of existing European or American firms is not clear. Krupp, for example, was active in Morocco from the late 1870s; there is evidence that Krupp made a small investment--not enough to exercise control but certainly enough to obtain information and play a role--in the stock of the German East Africa Company in the 1880s, presumably in the hope that minerals would be found in east Africa. French steel maker Creusot showed interest in Morocco in the early twentieth century, but this author has seen no evidence that it went further south.[24] In the 1920s, Union Carbide integrated backward, investing in mining manganese in the Gold Coast and chrome in Rhodesia.[25] By the post-World War II years, this type of supply-oriented investment by manufacturing firms in Africa would not be at all atypical, as major aluminum companies invested in bauxite and major steel companies went into iron ore in west Africa.[26]

It would be fascinating to explore the different problems of
technological transfer and diffusion when the European or American
multinational manufacturing company had expertise in the mining,
for example, the case of Krupp, where it had German iron mines,
and when the multinational company had no special know-how. Union
Carbide did not mine manganese in the United States; its investment
in mining in Africa was its first such stake. In the latter case, does
the multinational manufacturing company still have a technological
advantage over other nonmining firms in that it has access to information and the funds to buy technology and technically trained personnel and management? There would, however, seem to be a difference between the company that knows how to mine, already has
experience in mining and the particular technologies of mining, and
the firm that has never before carried forth this function and has to
obtain the technologies anew.

Interestingly, agricultural firms appear to have had similar
types of investment to those of the mining companies. Traders were
early investors in agriculture. Some found it satisfactory simply to
buy the crop produced by Africans. Others wanted to invest. The
expatriate-type investment in agriculture included companies floated
on European exchanges and those established by white settlers. To
some extent, the chartered companies embarked on agricultural
ventures.

European entrepreneurs and traders in the nineteenth and early
twentieth centuries started plantations. The Portuguese and Spanish
grew cocoa and coffee on Sao Tome and Fernando Po; the Belgians
planted rubber in the Congo; the British cultivated coffee, tea, and
cotton in Nyasaland--now Malawi--cotton, tobacco, and maize in
Rhodesia, maize in Kenya; the Germans had coffee, rubber, and
palm oil estates in Tanganyika, now Tanzania; the French, peanuts
in Senegal and bananas in French Guinea; and so it went. Some of
these early plantations survived into the interwar period. Many
were short-lived. New ones started in the 1920s. The British became active in sisal, coffee, and tea in Tanganyika and Kenya. They
also were active there in pyrethrum, which was a chrysanthemum
used for commercial purposes in the drug industry to treat certain
skin disorders; it was also early used as an insecticide. Indeed, of
the privately listed capital in Kenya in the 1930s, the largest part
was invested in agricultural estates, particularly in the sisal and
coffee plantations.[27]

Multinational agricultural giants such as United Fruit inaugu-
rated plantations in Africa. In 1934, in response to European bloc
trading during the depression, United Fruit made its first entry by
investing in French Africa. Operating through a French affiliate,
it developed plantations in the French Cameroons.[28]

International manufacturing enterprises also went into plantation agriculture in Africa. They invested as part of strategies of
backward integration, to obtain security of supply. They did not
want to be "squeezed by merchants or brokers," in Lever's case,
or by a cartel, in Firestone's case. Some investments were short-
lived, for example, Cadbury's in the Gold Coast, sold in 1912, and
Dunlop's in Liberia, sold in 1918. Probably the most important
multinational manufacturing enterprise to invest in African plantations was Lever Brothers. The parent company became Unilever
after 1929.

Early in 1902, William Hesketh Lever sent an investigator to
obtain information on potential raw material investments in west
Africa. The latter reported "an inexhaustible supply of Palm Oil
and Palm Kernels in the hinterland there only awaiting development
and the opening up of markets." Lever already had an interest in
two west African trading firms, but he wanted added control over
raw materials. First, and it would seem naturally, he looked to
British West Africa--in vain. The British Colonial Office believed
that in its West African territories, blacks should, in general, hold
their land as in years past and cultivate it as they wished. Lever,
by contrast, thought native methods miserably inefficient. He desired to introduce scientific cultivation of the oil palm and mechanical milling of its fruit. The Colonial Office offered Lever a 21-year
lease in Sierra Leone, which Lever scornfully rejected as entirely
inadequate.

Lever turned next to the Belgian Congo. The Belgians had a
different approach to colonial land policy and in 1911 Lever made a
formal agreement with the colony of the Belgian Congo that brought
into existence La Societe Anonyme des Huileries du Congo Belge.
The new company obtained rights to cultivate areas that totaled 1.8
million acres. By 1925, Lever's ventures in plantations and trade
were employing 25,000 Africans. [29]

Firestone Rubber Company was another multinational manufacturing firm that integrated backward to grow its basic raw material.
In the 1920s, Firestone started its immense rubber plantation in
Liberia. Its concession was for 1 million acres. By the 1960s, an
estimated 5 percent of the population of that country, 50,000 people,
lived on and were directly supported by the Firestone plantation. [30]

All these various mining and agricultural ventures transferred
to and within Africa more advanced techniques than had been present.
Mining and agricultural technology were transferred and to a limited
extent diffused, or adopted, by Africans. In the Gold Coast, skills
carried in by British mining firms were taught to Africans. Africans
learned how to operate power equipment. White developers of the
newly found Northern Rhodesian copper in the 1920s who had earlier

worked in the Gold Coast tell of the vast difference in skill level of their employees. [31] The mining companies diffused skills, production techniques, as well as the discipline of the industrial operation; they did not effectively diffuse the technology of entire, complex, mining processes. Where were the mining companies run by Africans, using western technology? Africans had neither the capital nor the managerial experience to develop their own similar ventures. While some agricultural know-how moved beyond the farm or plantation of European, or American, owners, much did not. Some European methods were adopted by Africans. For example, the plow was introduced to Africans in the eastern district of the Cape in the early nineteenth century. [32] It spread slowly northward. By 1926 some 26,000 plows were in use in Rhodesia; by 1936, 97,000. Not all of the 97,000 were employed by white settlers. [33] There was some reluctance by the British government to diffuse European technology. We have noted British policy in west Africa that opposed alienation of native lands and disruption of native methods of cultivation.* The Rhodesian secretary of agriculture in his report of 1930/31 stated that "No attempt should be made to induce changes in systems of native agriculture until the soundness of the improvement has been abundantly proved under the conditions in which the native farmer himself works."[34] On the other hand, sometimes there was conscious diffusion. Firestone, for example, taught Liberian independent growers how to plant rubber, giving them substantial technological assistance. The African growers never developed plantations to rival Firestone; their plots--some of which came to as much as 2,000 acres--were still dwarfed by the 1 million acre concession of Firestone (Firestone planted about 100,000 of its concession). [35]

In short, in the mining operations, Africans learned skills, specific to the jobs at hand. They learned industrial discipline. Not until just before, or sometimes after, independence did Africans begin to move into significant supervisory positions in advanced-technology mining operations. The technology linked with agriculture was, likewise, diffused only in a limited manner.

Indeed, the multinational companies that went to Africa to mine and to grow crops transferred western methods, succeeded very well in teaching limited skills and in training a work force, but proved far less adept in diffusing managerial or advanced technological methods. Since independence, however, governments have

----

*The policy was clearly not consistent. In the interwar years, the British colonial government did agree to lease to Lever Brothers plantations in Nigeria on 99-year leases. See Charles Wilson, Unilever, vol. 3, p. 213.

insisted on a change. Companies once operated by Europeans now
have Africans to a greater or lesser extent participating in management. The learning process goes forward slowly.

INFRASTRUCTURE INVESTMENTS

One of the most important contributions of foreign investors to
Africa lay in the building of infrastructure. Many of the investments
in railroads, roads, ports, telegraph facilities, and dams were undertaken by imperial governments. The foremost authority of the 1930s
on capital investment in Africa, S. Herbert Frankel, wrote that "minerals were the magnet which drew most of the existing railway lines
across the continent of Africa." He continued, "with the exception of
those in Rhodesia and Nyasaland, all the railways in British African
territories were constructed by, or on behalf of the Government concerned, and were owned and operated by it." There were public sector investments in railways in Belgian, French, Portuguese, and
German territories as well.[36]

Private enterprises also made investments. In French West
Africa, the Bordeaux firm of Maurel et Prom actively encouraged
the Niger railroad developments.[37] The Benguela Railroad went to
the port of Benguela in Angola from the copper belts of the Belgian
Congo and Northern Rhodesia. Although the greatest mileage on this
railroad was in Portuguese Angola and then in the Belgian Congo, the
financing was almost entirely by British private capital.[38] British
private capital was, after all, preeminent in Northern Rhodesian
mining and the railroad would bring the copper to the seacoast. In
Rhodesia, it was the chartered company, the British South Africa
Company, that built and financed the railroads; the investments were
directly linked with the mining activities of the British South Africa
Company. British capital also built the Beira Railway, the Beira-
Junction Railway, and the Trans-Zambesi Railway, sections of which
crossed the territory of Mozambique Company.[39]

Often infrastructure-type projects that were financed by an imperial government would actually be built by private, multinational,
construction firms. Thus, in the Sudan the Sennar Dam was constructed by a British firm, S. Pearson and Son, in 1921-25. The
dam was formally opened on January 21, 1926. The economist,
S. Herbert Frankel, has written that the occasion was "probably the
most important event in the history of the country since the recap-
ture of Khartoum." The dam opened the way for cotton production
on a major scale in the Sudan.[40]

When Lever Brothers and Firestone went to Africa, they found
that they had to provide their own railroads and roads within their

plantations. Lever Brothers laid railroad tracks in the Congo, while
Firestone put in roads in Liberia. Each developed communication
facilities. [41]

The infrastructure created by European firms clearly carried
new technology to Africa. Railroads opened up new territories and
the possibilities of development. Animal transport in tropical Africa
had been virtually impossible, because animals died of the prevalent
diseases. Railroads substituted for human carriers. A new technology of transport was brought to Africa, but to whose benefit? Likewise, to what extent did the cotton production in the Sudan serve
European purposes; to what extent African purposes?

Europeans clearly introduced, in a geographical sense, railroads, ports, roads, telegraph facilities, dams. But could Africans
build their own? Did they learn the technology of construction?
Here again, as in mining and agriculture, Africans learned certain
skills and work habits, but the management and the technology of the
entire processes long remained in European hands.

THE OIL COMPANIES

When American multinational oil companies first went to Africa,
they sought to sell their refined products. Vacuum Oil, part of the
Standard Oil group, had Africa as part of its territory. In 1885,
Vacuum Oil had established a London office. This office, in 1897,
sent two representatives to South Africa to explore the possibilities
of marketing lubricating oils and greases produced from petroleum.
In 1902, a branch was opened in Johannesburg. [42] (The product was
imported from the United States.) That same year, Vacuum Oil's
Paris company established a branch in Egypt. [43] As in many other
instances, the initial contacts of the foreign firm were in South Africa
and north Africa.

In 1907, the Lisbon office of Vacuum Oil negotiated with European trading firms to sell in west Africa. In Nigeria, for example,
Vacuum Oil, through British trading firms, supplied lubricants for
the palm oil presses and kerosene for lighting. Until June 1940,
Vacuum Oil, later Socony-Vacuum, and today Mobil Oil, supervised
its west African activities from Lisbon. [44] In 1940, the New York
office took over the supervision.

The predecessor of Mobil Oil appears to have been the first
multinational petroleum company to sell in Africa. Standard Oil of
New York, before it merged with Vacuum Oil, did some African
marketing, mainly in South Africa; Shell and the Texas Company followed; others came in later. A traveler to Africa in 1930 found
Vacuum products in Mauritania, Senegal, the Ivory Coast, Nigeria,

the Cameroons, the Belgian Congo, and Angola, "as well as in primitive villages in the continent's interior." In mining regions throughout Africa, there were Vacuum products. Likewise, the company had markets in Ethiopia and the Sudan.[45]

As for producing petroleum in Africa, as Babatunde Thomas' study (chapter 7) relates, there were early twentieth-century efforts made in exploration. For example, a German firm, the Nigerian Bitumen Corporation, began drilling in Nigeria in 1908.[46] Shell looked for oil in North Africa, finding some in Egypt before World War I.[47] In the 1920s, Sinclair sought oil, in vain, in Portuguese Angola and the Gold Coast.[48]

In 1938, a Shell subsidiary obtained a vast concession in Nigeria, to explore for petroleum. It found oil in commercial quantities in 1956.[49] Since the 1950s, it has begun to look as though there are substantial quantities of oil in Africa, in north Africa and also in west Africa, Nigeria, Angola--including Cabinda, Gabon, and elsewhere.

Since the 1950s, the functions of petroleum companies in Africa have changed substantially. Not only have companies become oil producers, but they are also oil refiners. Before World War II there were no oil refineries in black Africa. As independence came, new nations demanded that companies not only sell refined products, but actually refine within the host country. Refineries were built in Senegal, Kenya, Liberia, the Ivory Coast, and other countries that did not produce oil. Likewise, in nations that were new oil producers, there was an equal desire for refineries and petrochemical works. Companies built and made plans to build such manufacturing facilities. In Chapter 7 of this volume, Thomas discusses in detail the experience in Nigeria.

Thus, in recent years multinational oil companies have not only sold refined products and produced crude oil, but they have carried on manufacturing in Africa. They have introduced the most advanced technology. As Thomas' chapter in this volume points out, government revenues from the producing operations of multinational oil companies have provided funds for the government of Nigeria to commence the acquisition of new technology in sectors other than oil. In this case, the multinational oil companies are serving to diffuse technology, albeit indirectly.

MULTINATIONAL MANUFACTURING COMPANIES

Multinational manufacturing companies, like the multinational oil enterprises, went first to Africa to sell. Some established sales outlets. Before World War II, if the multinationals manufactured in

Africa, it would nearly always be in South Africa. The reason was
that the markets in black Africa were not large enough to warrant
manufacturing. Neither tariff barriers nor competitive pressures
compelled it. Multinational manufacturing companies also went to
Africa for supplies. As we have seen, some integrated backward to
invest in mining or agriculture.

To the extent that the manufacturing company was a trader, it
introduced new wants, as did other traders. We have discussed its
role as investor in mining and agriculture. Sometimes, in connection with mining and agriculture, companies built processing plants.
Not until the late 1950s and 1960s did multinational manufacturing
enterprises begin on a general basis to establish subsidiaries with
factories to serve African markets. When they did so, it was because they could no longer penetrate those markets with exports
from abroad. From pharmaceuticals to detergents, products once
imported were made in Africa.

The change during the postwar period was dramatic. An example lies in the transformation of Unilever's Nigerian operations
from 1949 to 1965. In 1949 Unilever, through United Africa Company, grew bananas, oil palm, timber, and rubber in Nigeria. By
1965, the company manufactured in Nigeria rubber cushions and
pillows, crepe rubber and sheet, vehicles (assembly only), leaf
springs, cigarettes, meat products, packing materials, cotton yarn,
plastics, prestressed concrete, cement, bicycles, mattresses, furni-
ture, radios (assembly only), sewing thread, cotton textiles, ice
cream, car batteries, toilet preparations, stout, beer, mineral
waters, reconstituted milk, and stationery. By 1965, Unilever had
invested in 72 factories in tropical Africa, sometimes on its own
and sometimes in joint ventures, mainly with European partners.
It provided commercial management for 55 of these ventures and
technical management for 34.[50]

Whether a multinational company established a processing
plant in connection with its mining or agricultural venture or a manufacturing unit to serve host nation markets, the multinational corporations transferred technology. The diffusion process was, however,
slow. It took time for Africans to learn the foreign technology.
United Africa Company, the Unilever subsidiary, had established its
first training school in Nigeria in 1934, at Burutu, which was the
headquarters for the company's river fleet. The technology taught
there was not that of manufacturing. At Sapele, Nigeria, where it
had a timber operation, and came to manufacture plywood and block-
board as well as crepe rubber and sheet, United Africa Company established another training school. One of its partners in manufacturing, Niger Motors, opened a training school at Kano, in Northern
Nigeria.

United Africa Company also started training schools in Ghana and in east Africa, specializing in mechanical technology. Africans were sent by the company to Europe for technical training with firms such as Guinness, Vauxhall, and Heinekens.[51] Other companies followed similar patterns. Clearly, diffusion was proceeding. Africans with degrees in engineering were learning processes that were imported into their nations. The basis for assimilation of the technology was established. Recently, Africans have begun to insist on ownership participation in the new manufacturing ventures, as Thomas has indicated in his introduction to this volume.

CONCLUSION

In conclusion, the waves of foreign companies bringing technology from overseas have acted to alter the economies of the African continent. While the pace of what has happened in Africa differs from elsewhere, while the specifics are unique, it seems evident that multinational companies carrying on trade, mining, agriculture, construction of infrastructure-type facilities, as well as manufacturing, have moved worldwide, and Africa has been included. Technology has been transferred to Africa, that is, physically carried there, in terms of products, concepts, and processes. It also has been absorbed to varying extents. The absorption has been a slow and gradual affair. The results, however, should not be underestimated. The change has been substantial.

The first absorptions were clearly of product rather than processes--and were of new crops. The second absorptions were more conceptual than concrete, of production for market of particular crops, of desires for particular goods. The third absorptions were of processes--of specialized skills, of particular equipment, from plows to shovels, and of the discipline of modern production. The fourth absorptions have been most recent and are still occurring-- absorptions of more complex and sophisticated processes. In a sense, the earlier absorptions provided the foundations for what is taking place today.

When foreign companies established new productive facilities in Africa and introduced modern mining, agriculture, and industry, sometimes the foreign company already had within its enterprise-- that is, either it or its affiliates had or knew--the technology. Thus, when American Metal Company mined in Northern Rhodesia, it had experience elsewhere; when United Fruit went to the Cameroons, it had long had banana plantations in Central America and the Caribbean; and when Shell-BP built a refinery in Nigeria, its parent firms had built many refineries before. On the other hand, it is clear that when many foreign "expatriate" investors went to mine, when companies such as Union Carbide integrated backward to mine, when Firestone grew rubber, and when Unilever embarked on certain manufacturing in Africa, each had to acquire the basic technology from outsiders. We need to know more about the relationship between ownership and technological transfer and diffusion. Does a western firm have better access than an indigenous one to the required technology, even though it doesn't have it within its own organization? Was this true in the past, but not true now? Does any particular type of corporate ownership relationship lead to greater technological diffusion?

Before the 1950s, foreign companies that went to Africa, and elsewhere, diffused technology primarily because it was profitable for them to do so. New crops were introduced to provide a basis for profitable trade. New concepts offered the basis of profits. The teaching of specialized skills meant a better work force or cheaper, higher quality, purchased supplies and services. Industrial discipline meant higher productivity. Sometimes it was politically desirable, and thus profitable, for the foreign firm to diffuse technology. The money made by independent rubber producers in Liberia created friends for Firestone; the independents were not large enough to provide competition. The multinational corporation in the post-World War II period, as earlier, has operated with a goal of private profit. Multinational corporations are not charities or public service enterprises. They have technological know-how and talent and can, will, and do transfer and diffuse this technology within the context of private gain. They often have superior access to others' technology, and it will usually be easier for them to acquire new technology than for a firm in a developing country. This is because they have the capital and trained technical personnel to do so. If it is economically in the best interest of the corporation, and the economics can be changed by political considerations, then, we would like to suggest, the corporation becomes an excellent diffuser of technology. There may be questions as to whether the multinational corporation is diffusing the right technology, the one appropriate in terms of each nation's factor costs or in terms of a country's need to create full employment. This is, indeed, an important issue. There also may be questions on the cost of that technology. This chapter only argues that if it is economically in the best interest of the corporation, as perceived by the corporation, then the corporation serves as an excellent diffuser of the technology, especially the technology that it knows and uses.

On the other hand, if the economic gain to the corporation is not evident, or if to diffuse technology means to create competition without commensurate rewards, or if by altering its technology to meet appropriate factor costs for engineering or information cost reasons the profits will not be maintained, then the multinational corporation would seem to be a less effective diffuser of technology. Nonetheless, often there are circumstances wherein the multinational corporation transfers technology, but plays a passive or negative role in the diffusion of the technology, yet the diffusion still occurs: competitors hire men trained by the corporation to the corporation's dismay; competitors imitate the processes developed by the corporation to the corporation's disenchantment. Much depends on how the receiver of the technology is able to digest what the multinational firm transfers.

In short, multinational investors have transferred to Africa over the years a vast range of technologies, incorporated in products, including crops; concepts, including a wide range of attitudes; and production methods, from plows to oil refineries. Some of the technologies were diffused, that is effectively absorbed. In all cases, there was a process, from the physical transfer to the diffusion, adoption by Africans; often, the process was a slow one; in no two cases was the process identical; diffusion of each product, concept, and production method took a different path. There remain cases of transfer without diffusion. This does not mean that diffusion, in time, will not occur. Over the years there has been substantial diffusion. More seems inevitable. What needs further study is not how the foreign investor transfers or carries the technology over borders, but rather what are the conditions whereby there is absorption of the transferred technology within the recipient country, what are the time lags, and how can these conditions be improved and the time lags shortened by public policies.

NOTES

1. These data are from material the present author assembled for Richard Morris and Graham Irwin, eds., Encyclopedia of Modern World History (New York: Harper and Row, 1970), p. 673. See also Fernand Braudel, Capitalism and Material Life 1400-1800 (New York: Harper and Row, 1974), p. 70.

2. Robert I. Rotberg, A Political History of Tropical Africa (New York: Harcourt, Brace and World, 1965), pp. 139-42.

3. Ibid., p. 158.

4. S. Herbert Frankel, Capital Investment in Africa (London: Oxford University Press, 1938), pp. 22-23.

5. Ibid., p. 65.

6. Data from New York Life Insurance Company.

7. Frankel, op. cit., p. 53.

8. Ibid. , p. 61.

9. Ibid. , pp. 61n. , 63-64.

10. Ibid. , pp. 61, 63, 68, 74; George T. Kimble, Tropical Africa, 2 vols (New York: Twentieth Century Fund, 1960), vol. 1, p. 70; private interviews.

11. This author has learned a great deal from correspondence with John Stopford. See also his "The Origins of British-Based Multinational Manufacturing Enterprises," Business History Review 48 (1974): 305-06.

12. See Herbert Feis, Europe: The World's Banker 1870-1914 (New Haven: Yale University Press, 1930), p. 15n.

13. See also Mira Wilkins, "Multinational Enterprises," in The Rise of Managerial Capitalism, ed. Herman Daems and Herman van der Wee (The Hague: Martinus Nijhoff, 1974), p. 217.

14. Frankel, op. cit. , p. 162.

15. Ibid. , p. 79.

16. Mira Wilkins, The Maturing of Multinational Enterprise: American Business Abroad from 1914 to 1970 (Cambridge, Mass.: Harvard University Press, 1974), pp. 111-12.

17. Feis, op. cit. , pp. 254-55; Theodore Gregory, Ernest Oppenheimer (Cape Town: Oxford University Press, 1962).

18. Frankel, op. cit. , p. 292.

19. Private interviews.

20. D. K. Fieldhouse, Economics and Empire 1830-1914 (Ithaca: Cornell University Press, 1973), pp. 350-52.

21. Societe Internationale Forestiere et Minere du Congo, Statuts (Brussels: Societe Internationale Forestiere et Minere du Congo, 1950), and data from Albert Van de Maele, Guggenheim Bros. , New York, April 19, 1965.

22. Wilkins, Maturing, op. cit. , p. 110.

23. Frankel, op. cit. , p. 314.

24. Fieldhouse, op. cit. , pp. 286, 302, 373.

25. Wilkins, Maturing, op. cit. , pp. 112-13.

26. Kaiser Aluminum, U.S. Steel, and Republic Steel were among the many aluminum and steel companies that invested in west Africa. See Andrew M. Kamarck, The Economics of African Development (New York: Praeger, 1967), p. 146.

27. Kimble, op. cit. , I, 145-46; Frankel, op. cit. , p. 163.

28. Wilkins, Maturing, op. cit. , p. 196.

29. Stopford, op. cit. , p. 32 (on Cadbury and Dunlop). Charles Wilson, Unilever, 3 vols. (New York: Praeger, 1968), I, 159, 165ff, 290; and Kamarck, op. cit. , p. 191 (on Unilever).

30. Wilkins, Maturing, op. cit. , pp. 99-102; and interview with A. Lotz, manager, Firestone Company, Liberia, July 16, 1965.

31. Interviews in Rhodesia and Zambia, 1965.

32. Morris and Irwin, op. cit., p. 673.

33. Frankel, op. cit., p. 244n.

34. Quoted in ibid.

35. Interview with Romeo Horton, Monrovia, Liberia, July 15, 1965.

36. Frankel, op. cit., pp. 374, 377, 407-19.

37. Fieldhouse, op. cit., p. 317.

38. Frankel, op. cit., pp. 169, 214, 375, 416.

39. Ibid., pp. 377ff, 416.

40. Ibid., p. 362.

41. Wilson, op. cit., p. 178; interviews; personal observations.

42. Data from Mobil Oil Southern Africa (PTY) Ltd., "History of the Company," typescript, ca. 1955.

43. Mimeographed data on Mobil Oil Egypt, S.A., n.d.

44. S. J. Lofting, "Mobil in Nigeria--A Friend for More than Fifty Years," unpublished typescript, June 7, 1954. Obtained from Mobil in Lagos, Nigeria.

45. Socony Mobil Oil Company, Inc., "A History in Brief," pamphlet published by company, n.d., p. 32.

46. See chapter 7 of this book.

47. Data from interviews in Cairo with David Washburn, Mobil Oil, August 21, 1965.

48. Wilkins, Maturing, op. cit., p. 121.

49. See chapter 7 of this book.

50. Wilson, op. cit., vol. 3, pp. 222-23, 225.

51. Ibid., p. 224.

3

FOREIGN INVESTMENT POLICY, PUBLIC INVESTMENT IN INDUSTRY, AND NATIONAL SCIENCE AND TECHNOLOGY POLICY

INTRODUCTION

One important general conclusion to be drawn from Chapter 1 concerning the flow of technological know-how from the highly industrialized countries to African countries through DFI is that the viewpoint and the absorptive capabilities of the individual recipient countries are often guessed, derived theoretically, or neglected. Until the mid-1960s, two critical entities, namely the government policy makers and the indigenous scientific communities in African countries, have played only passive roles in the acquisition and adaptation of foreign technology.

This chapter analyzes changes, since the mid-1960s, in public policies on DFI and on the development of science and technology, S and T. Policy changes affecting these two areas are being instituted by an increasing number of African countries, primarily to encourage and promote active roles in the national economy for the two entities mentioned above and for indigenous entrepreneurs as well. One of the major changes in public policy on DFI already examined in Chapter 1 is the increasing use of specific instruments, investment codes in particular, to ensure selectivity, effective monitoring, and control by host governments over DFI. The initial effect of this particular shift in public policy has been the rising number and volume of equity participation by host governments in foreign businesses, usually large-scale industries, for immediate local majority controlling interest or eventual complete ownership. For example, Nigeria has increased its equity participation in the oil industry, demanded and obtained joint ventures with DFI in tanker, petrochemical, liquefied natural gas, and steel investment projects; and Zaire has gained participation and ownership in the development

of its phosphates by DFI. The wave of expansion in domestic public
equity participation in private foreign enterprises, however, has
tended to be haphazard. There also is a general lack of due consideration for fundamental technological requirements that are necessary for the real domestic control of the private sector. The general
pattern of public investment in the private sector and its various
meanings are investigated in this chapter. The development of S and
T policy also is examined to ascertain the readiness of a number of
African countries to fill the void that is likely to result from the
shift and probable decline in the future participation of DFI. Some
of these shifts already have begun to take place primarily in Nigeria
and Ghana. Both countries instituted major changes in their investment policies, in 1973 and 1975 respectively, by the indigenization
of equity and management control in foreign companies, covering a
substantial part of the private sector. These policy changes invariably are intended to reduce the relative levels of economic dependence
especially on DFI, and promote structural changes for long-term
development in these countries. However, apart from its obvious
political implications, there are major technological implications for
the desired improvements in their relative economic independence.
Clearly, the success of this policy in all its various conceptualiza-
tions is dependent on the extent to which these countries have the
requisite scientific, technological and entrepreneurial manpower,
and the availability of financial resources for ongoing development
and maintenance of the implied infrastructure. The manner in which
attention has been directed toward the gradual fulfillment of these
requirements is examined below. First, this is preceded by a brief
analysis of foreign investment policy and its implications for the importation of appropriate technological know-how.

FOREIGN INVESTMENT POLICY, PUBLIC INVESTMENT IN INDUSTRY, AND THE IMPORTATION OF TECHNOLOGICAL INNOVATIONS

    Diffidence about keeping DFI in check in African countries, and
the strong national desire for indigenous control of key resources as
well as sectors considered vital to national economic development,
have necessitated major changes in foreign investment policy. The
new changes seek to monitor the flow of foreign investment, diversify
the sources of investment funds, negotiate better terms for the host
countries, increase local participation in the growth of the private
sector, minimize and, where deemed necessary, phase out the past
active roles by foreign entrepreneurs.

Although there is no outright prohibition of foreign investment
in African countries, and most still have liberal investment codes to
attract foreign capital, there is a definite shift in the flow of DFI
away from smaller countries to the large and relatively well-endowed
countries like Nigeria, Zaire, Zambia, and Ghana in spite of stringent codes in these countries' foreign investment policies. Some of
the factors to which this trend has been attributed are discussed in
Chapter 4.

As illustrated in Chapter 1, India's, Mexico's, and the Andean
group's approaches to foreign investment continue to serve as models
for formalized and coordinated approaches to the control of DFI by
many DCs. The Andean group explicitly attempts to acquire technology on the most advantageous terms by tying the inflow of appropriate
technology to the inflow of DFI. The group requires 51 percent local
ownership as a minimum in new ventures involving DFI in member
countries, and complete local ownership is also required after 15
years in operation. One of the unique characteristics of this approach is the opportunity for the preclusion of the usual advantage
that foreign investors have in being able to play one country against
another in negotiating for highly favorable terms. There is implied
causation between the group's investment policy, specifically the
mandated complete acquisition of foreign establishments by domestic
investors within a prespecified period, and the observed decline in
DFI since the policy was put in force, especially in the last two
years. This apparent relationship, which has yet to be empirically
verified, raises a major question about the extent to which the negative reactions by foreign investors reflect their long-term orientation in the DCs, and thereby dispels their presumed primary desires
for profit maximization in the short-run.

Contrary to the expectation of policy makers in many African
countries, the obvious political and economic necessities of rapidly
phasing out the economic dominance of DFI and the associated foreign
entrepreneurship have left voids that must be filled by the increasingly active government role through the growing number of public
enterprises. Despite increases in public, and to a minor extent
private, local equity participation, the real rather than the apparent
control of the affected foreign business operations remains in expatriate hands. The general expectation that local equity participation necessarily ensures host-country orientation by the subsidiaries
of foreign firms has turned out to be an illusion. Although the level
of entrepreneurial know-how in several African countries is more
advanced than is usually acknowledged, the prevailing nonconcurrence
of equity-ownership and control, suggests that the indigenization and
local control of the private sector are more likely to be attained
through the development of the requisite technological and entrepre-

neurial manpower and its integration into a time-phased national
manpower development program.

In case studies of complete acquisition or expropriation of foreign firms that have necessitated the replacement of expatriate personnel, the general lack of scientific, technological, and entrepreneurial infrastructure has tended to obviate such sweeping and rapid
indigenization programs. In 1967, for example, Zaire, formerly the
Belgian Congo, chose to diversify DFI in its mining industry, including copper, diamonds, and cobalt, which has been dominated by
Societe Generale, a holding company of Union Miniere. The latter
was for a long time the dominant industrial influence in the mining
province of Katanga. The diversification efforts failed because of
the imminent mass exodus of foreign personnel in protest, and the
obvious recognition that the country was ill-prepared for their replacement by local personnel. The government modified its plan
and, under a relatively flexible arrangement, local personnel began
to undergo appropriate training for the eventual replacement of the
Belgians. Apart from the modification in policy, major public investment was also undertaken to commence the development of the
requisite technological and entrepreneurial manpower. Zambia and
Nigeria have also had similar experiences in their copper and oil
industries respectively. One of the lessons from these pervasive
experiences in DCs is the recognition of the need for appropriate
policy formulation for scientific and technological development in the
search for the desired and practical level of economic independence
by an increasing number of these countries.

NATIONAL SCIENCE AND TECHNOLOGY POLICY AND
LONG-TERM TECHNOLOGICAL DEVELOPMENT

The development of national science and technology policy in
African countries is of relatively recent origin, dating back only to
the mid-1960s. Its process and general nature suggest a definite
recognition of the minimal technological contributions from DFI, the
need to develop a road map for the acquisition of foreign technology,
and the acceptance of the necessity for national commitments to the
development of indigenous scientific and technological manpower. The
courses of policy changes and their implications for long-term technological and economic development are examined variously in the
following pages.

The scarcity of financial and material resources for scientific
and technological development of African countries, constitutes a
major constraint on the indigenization programs outlined above, and
on national development. The efficient development and utilization of

S and T as instruments of economic, social, and political development also require policy formulation, planning, and coordination.
National science and technology policy is an expression of the degree
of national scientific consciousness and the recognition of its technological manifestations in the national economic development process. The formulation of a national S and T policy usually consists
of legislative and executive measures and the development of appropriate policy tools for its implementation. The purposes of these
tools are to encourage scientific research, to ensure that the research fit into the medium- and long-term national development plan
in a manner that the foreign inflow of technology, and the national
scientific and technological potential are fully exploited for the
achievement of national development objectives. In formulating
science policy, it is therefore essential to undertake, first an evaluation of the available scientific and technological related resources--
such as human, and material resources; second, a dynamic evaluation of human resources as vital input for applied research in educational institutions, industry, and the public sector; and last, a periodic monitor of the proportion of the actively employed and total
population accounted for as the scientific and technological manpower.
    The objective of the new S and T policies in various African
countries is to focus on these factors and specifically to minimize
existing shortages--quantity, and more importantly, the quality--of
scientific and technological manpower at all skill levels. Furthermore, through these policies, new channels and modes of transfer
are sought for the acquisition, facile assimilation, advantageous use
of appropriate imported technology, indigenous human resources, and
optimal exploitation of national raw materials locally. Although the
implications of skilled manpower deficiencies for economic and social development efforts vary from one country to another, one other
common theme seems to prevail; that is, to derive maximum benefits
from the fruits of science and technology for improvements in society's welfare.
    The 1964 UNESCO Conference in Lagos, Nigeria on the "Or-
ganization of Research and Training in Africa," represented the
initial recognition of the pressing need for national science and technology policy in the technological development of African countries.
Before that conference, the majority of African countries had developed neither a national science and technology policy nor a national
institution to coordinate scientific research activities.
    The Lagos Plan, as the outcome of the Conference was later
known, sought to encourage and promote national S and T policy, and
the long-term plan for its implementation in individual countries.
But the Conference left unanswered the question of what role multilateral arrangements should play in African scientific and technological

manpower development. A follow-up on the 1964 Conference was
held in Yaounde, Cameroon in 1967. From both conferences four
major impediments to scientific and technological development in
African countries were identified. These were absence of adequate
national scientific policy; lack of any machinery for coordinating and
preparing such policy; extreme shortages of senior scientific and technical personnel; and lack of enough training and research institutes. [1]

Among the recommendations proposed by the 1967 Conference
for national policy actions to overcome these impediments were a
minimum allotment of 0.5 percent of GNP to research, and an estimated 200 scientific workers per million of the population should be
the primary targets all African countries ought to attain by 1980.

Table 3.1 shows the position of a representative number of
African countries for various years during the 1960s and the early
1970s in terms of their financial expenditure for the promotion of
S and T, and their estimated supply of scientific and technological
manpower. Some of the data presented in this table are estimates
and require caution in their interpretations. In the case of the Ivory
Coast, for example, 1.6 billion francs CFA in national expenditure
on R and D as of 1967 seemed an impressive allocation for 378 scientific and technical personnel when compared with other African
countries; but 806 million francs CFA of that expenditure was accounted for by foreign aid.

Estimates of the proposed targets on the two recommendations
above are presented for 1970, 1975, and 1980 in Table 3.2. These
recommendations demonstrate the lack of primary consideration for
existing fundamental differences among these countries. A few
countries, notably Nigeria, Ghana, and Senegal, had already surpassed these targets by the late 1960s. For example, from 1971
through 1973, estimates of recurrent expenditures on scientific research in Nigeria averaged approximately 1 percent of the annual
gross national product. Since the late 1960s, Senegal has been
spending in excess of 0.7 percent of its GNP annually on scientific
research. By 1971, Ghana, Malawi, Tanzania, Mauritius, and
Nigeria had surpassed the targeted 200 indigenous scientists and
engineers per million population (see Table 3.3). One of the primary aims of the Nigerian government in the 1973/74 budget is a
review of the existing procedures for remitting royalties and other
payments for patents, management, and technical fees. There is a
definite recognition of the need for improvements in these modes of
payments and the corresponding channels for the transfer of technology in some of the other African countries, prominently Cameroon,
Guinea, Kenya, Ethiopia, Zambia, and Zaire. Although there is
insufficient basis for the comparison of the early 1970s experiences
with conditions that prevailed in the 1960s for these countries, there

TABLE 3.1

Cumulative Financial and Human Resources in Science and Technology for African Countries in the Mid-1960s and Early 1970s

Country	Year	Scientists and Engineers			Technicians			National Expenditure on Research and Development*
		Total (1)	Engaged on R and D (2)	2 as Percent of 1	Total (3)	Engaged on R and D (4)	4 as Percent of 3	
Botswana	1967	180	10	5.6	606	6	.9	317,500 rand[a]
Cameroon	1967	800	80	10.0	3,000	140	4.6	350,000,000 francs CFA (159,399,090)
Central African Republic	1969	124	--	--	--	--	--	
Congo (Brazzaville)	1966	231	17	7.4	415	3	.7	248,300,000 francs CFA (113,082,235)
Ethiopia	1967	4,406	--	--	7,753	--	--	--
Ghana	1966	5,137	167	3.3	27,893	551	2.0	--
	1970	6,897	--	--	15,096	--	--	1,661,000,000 francs CFA[b] (756,465,000)
Ivory Coast	1967	--	204	--	--	174	--	--
Kenya	1964	2,536	--	--	6,893	--	--	--
	1970	3,000	--	--	8,900	--	--	--
Malagasy	1966	--	195	--	--	230	--	997,000,000 francs MG
Malawi	1967	1,253	14	1.1	5,028	--	--	--[c]
Mauritius	1967	554	--	--	2,010	--	--	--
Nigeria	1966	3,970	1,723	43.4	6,997	--	--	N9,770,000 (13,475,000)
	1970/71	19,885	--	--	15,241	--	--	
Rwanda	1967	207	19	9.2	454	7	1.5	24,400,000 francs Rwanda
Somalia	1965	175	4	2.3	938	10	1.1	770,000 Somali shillings[d]
Tanzania	1968/69	4,080	--	--	10,943	--	--	
Togo	1967	215	--	--	--	--	--	18,108,000 francs CFA (9,002,350)
	1971	461	--	--	388	--	--	
Upper Volta	1967	160	48	3.0	102	32	31.4	

*Approximate dollar equivalents are given in brackets.
[a]Provisional.
[b]Of which 806 million francs CFA is foreign aid.
[c]Total expenditure by the government (of research) is 0.3 percent–0.4 percent of GNP.
[d]For 1967 only.

Sources: Compiled from The Promotion of Scientific Activity in Tropical Africa. UNESCO, Document NS/SPS/11, SC/SP. 68. X111.11/A Paris, 1969, p. 21, National Science Policies in Africa: Situation and Outlook. Studies and Documents No. 31, UNESCO 75700 Paris, France, 1974, p. 88.

is evidence of significant increases in budgetary allocations for scientific and technical training, and research and development since 1969 (see Table 3.4).

TABLE 3.2

Target Number of Scientists Needed as a Function of Projected Population Increases and Interrelationship Between Gross National Product and Research Expenditure Targets for African Countries, 1970-80

	1970	1975	1980
Population (in thousands)	277,882	312,333	353,243
Aggregate number of scientists at 200 per million population (thousands)	55.6	62.4	70.6
Gross national product (millions of U.S. dollars)	39,696	49,895	65,238
Expenditure for education (in millions of U.S. dollars)			
Total expenditure	2,865.0	3,492.6	4,160.4
Higher education	499.6	722.6	1,029.7
Expenditure for research (in millions of U.S. dollars)			
Total expenditure	198.5	280.6	326.2
Fundamental research	39.7	56.1	65.3
Other research	158.8	224.5	260.9

Source: Outline for a Plan for Scientific Research and Training in Africa. UNESCO in association with UN-ECA. Paris-7e 1964, p. 21.

An examination of developments in numerous other African countries suggests that the two targets above have been taken seriously, but targets are meaningless when misconstrued as ends in themselves as it has been the case notably in the countries identified above. The overall response to the shortage of scientific and technological manpower has led to duplicative, rather than innovative, manpower development policy and planning. In the next section, the basic institutions for the development of scientific and technological manpower are examined and the nature of the link between them and the primary ultimate users of this manpower, namely industry and government, is examined.

TABLE 3.3

Human Resources in Science and Technology

Country	Year	Estimated Mid-year Population (millions)	Scientists and Engineers	Technicians	Total Scientists, Engineers, and Technicians	Number of Technicians per Scientist and Engineer	Scientists, Engineers, and Technicians per Million Inhabitants	Scientists and Engineers per Million Inhabitants	Estimated Economically Active Population (millions)	Scientists, Engineers, and Technicians per Million Economically Active Population
Cameroon	1967	5.52	800[a]	3,000	3,800	3.8	688	145	2.26	1,680
Central African Republic	1969	1.58	124[b]	--	--	--	--	80	--	--
Congo	1966	0.86	231	415	646	1.8	751	270	0.32	2,030
Ghana	1970	8.64	6,897	15,096	21,993	2.2	2,545	795	3.33	6,600
Kenya	1970	11.25	3,000	8,900	11,900	3.0	1,058	270	3.60	3,300
Malawi	1967	4.12	1,253	5,028	6,281	4.0	1,524	300	1.36	4,620
Mauritius	1967	0.77	554	2,010	2,564	3.6	3,330	720	0.22	11,900
Nigeria	1970/71	55.07	19,885	15,241	35,126	0.8	638	360	22.03	1,595
Rwanda	1967	3.42	207	454	661	2.2	193	60	2.00	333
Tanzania	1968/69	12.59	4,080	10,943[c]	15,023	2.7	1,193	320	5.80	2,600
Togo	1971	2.02	461	211	672	0.5	333	230	0.73	925
Upper Volta	1967	5.05	160	102	262	0.6	52	32	2.70	98

[a]Estimate.
[b]Partial data.
[c]Data relate to persons employed in any job normally requiring 1 to 3 years formal post-secondary education/training.

Source: UNESCO, National Science Policies in Africa: Situation and Future Outlook, Science Policy Studies and Documents No. 31, UNESCO 75700, Paris, France, 1974, p. 88.

TABLE 3.4

Total and Current Expenditure for Research and Experimental Development

Country	Fiscal Year Beginning	Currency	Total Expenditure (in thousands)	Current Expenditure (in thousands)	Current Expenditure as Percent of Total	R & D Expenditure as Percent of GNP (estimates)
Algeria	1972	Dinar	78,000	68,000	87	0.4
Cameroon	1967	Franc CFA	350,000	--	--	0.2
Central African Republic	1969	Franc CFA	183,908	175,971	96	0.3
Chad	1969	Franc CFA	227,060	221,360	97	0.4
Congo, People's Republic of	1966	Franc CFA	--	248,322	--	0.4
Gabon	1970	Franc CFA	1,895	1,882	99	0.0
Ghana	1971	New cedi	--	21,612	--	0.8
Ivory Coast	1970	Franc CFA	1,401,124	1,385,124	99	0.4
Kenya	1971	Shilling	102,940*	83,160*	81*	0.8
Madagascar	1971	Franc	2,294,000	2,294,000	100	0.9
Malawi	1971	Kwacha	1,581*	1,418*	90*	0.5
Mauritius and deps'.	1970	Rupee	5,124*	4,290*	84*	0.5
Nigeria	1970	Pound	11,900	11,000*	92*	0.5
Rwanda	1967	Franc	--	24,400	--	0.1
Senegal	1971	Franc CFA	--	2,176,000	--	1.0
Somalia	1967	Shilling	--	770*	--	--
Sudan	1971/2	Pound	--	3,323	--	0.6
Togo	1971	Franc CFA	--	1,070,829	--	1.1
Tunisia	1971	Dinar	3,000*	--	--	0.4
Upper Volta	1970	Franc CFA	412,768*	--	--	0.5

*Estimates.

Source: UNESCO, National Science Policies in Africa: Situation and Future Outlook, Science Policy Studies and Documents No. 31 UNESCO 75700 (Paris, 1974), p. 96.

SCIENTIFIC AND TECHNOLOGICAL MANPOWER DEVELOPMENT,
NATIONAL RESEARCH INSTITUTIONS, AND OPTIMAL
UTILIZATION OF SCIENCE AND TECHNOLOGY FOR
DEVELOPMENT PURPOSES

A basic prerequisite for the development of scientific and technological manpower, and hence the development and efficient use of
research institutes, is the development of appropriate skills through
technical and higher educational training. The scarcity of scientific
and technological manpower in most parts of Africa is a manifestation of the short supply of institutions necessary for its development,
namely, universities, colleges of science and technology, intermediate level technical training schools and trade schools. Other deficiencies include a lack of coordination of the activities of these institutions, and a general lack of explicit national manpower development
planning.

In 1969, there were about 33 universities in African countries
and most had faculties of science, engineering, and technology, but
their research activities were rarely directly applicable to manifest
and planned national needs. Since 1970, the number of these institutions has increased considerably. Table 3.5 shows the pattern of
changes in enrollments since 1960 by subregions based on incomplete
ECA estimates.[2] From the late 1960s through the early 1970s, enrollments at all levels of education in African countries grew at the
rate of 4.3 percent in the west, 6.4 percent in the east, and 8.6 percent in central Africa. When the period since the mid-1960s is compared with the immediately preceding decade, there is evidence that
the rate of growth in enrollments has slowed as resources devoted to
educational and other forms of training fail to keep pace with appropriate manpower development needs because of shortages of qualified
teachers and limited budget. For example, the average rate of
growth in secondary and higher education slowed considerably from
13.4 and 13 percent during the 1960-65 period to 7 and 6 percent
during the period since 1965-68 respectively. However, two sub-
regions--east and central Africa--experienced continued growth in
enrollments.

Research activities in and out of universities are limited to
eight broad fields of activity (see Table 3.6). Food and agricultural
sciences account for the majority of personnel and number of institutions, but not necessarily in budget allocations. The conduct of industrial research has also been limited. This is due primarily to
scientific and technological manpower constraints, ambiguous national
priorities, and nonhost country-oriented directives from multilateral
and bilateral aid-giving agencies. Until the late 1960s, the percep-
tion of these agencies by virtue of the heavy dependence on their

financial support has been that they possess all the wisdom on the
strategy for long-term development in the DCs. It is now generally
recognized that this judgment is erroneous.

TABLE 3.5

Higher Educational Enrollments by
Subregion of Africa, 1960-70

	North Africa	West Africa	Central Africa	East Africa	Total
Estimated enrollments (in thousands)					
1960	131.2	13.0	5.2	10.0	159.4
1965	217.2	23.9	8.2	16.0	265.3
1967	218.8	25.4	12.6	23.4	280.2
1968	226.6	27.9	15.2	26.0	295.7
1969	251.4	29.6	17.2	30.7	328.9
1970	270.0	32.0	19.7	33.6	355.3

Source: Adapted from United Nations, Survey of Economic
Conditions in Africa, 1971 Part I (New York: United Nations, 1972),
pp. 32-33.

In infrequent instances where research institutions other than
those in the institutions of higher learning have endeavored to fill the
void in industrial research, the virtual lack of policy formulation and
coordination has tended to impede their effectiveness. These institutions remain relatively ineffective, but their number also has
grown considerably.
The results of two UNESCO surveys conducted in 1963-64 and
1969-70, that is, between the 1964 Lagos Conference and the 1967
Yaounde Conference, show that scientific and technical institutions
engaged in one form of research or the other in tropical Africa has
proliferated to over 2,000 with a research work force of about
11,000.[3] See Table 3.6.
The increases, which were accounted for primarily by a significant use of legislative instruments rather than anticipated or
planned effective demand by industry, were particularly substantial
in the east and west African subregion; and in instances of some countries, namely Gambia, Liberia, Niger, Zaire, Tanzania, and
Zambia, the number of national research institutions, NRIs, more
than doubled.

TABLE 3.6

Number of Research Institutions Engaged in the Main Fields
of Activity in 1964 and 1969 in 40 African Countries

Fields of Activity	Number of Institutions[a]		Total Indigenous Research Workers	
	1964	1969	1964	1969/70
Fundamental sciences	205	392	406	2,298
Earth and space sciences	239	256	737	1,629
Medical sciences	114	163	435	1,907
Food and agricultural sciences	581	484[b]	1,773	3,709
Research on fuel and power	8	29	6	149
Industrial research	37	107	71	962
Economics	--	40	--	97
Social and human sciences	--	94	--	342

[a]Total number of research institutions is less than the sum of
the numbers shown in the table because many institutions carry out
research in several fields.

[b]This figure does not show the complete number of institutions
carrying out research in the field of food and agriculture.

Sources: UNESCO, Scientific Research in Africa. National
Policies, Research Institutions (in association with ECA), 1966;
UNESCO, Survey of the Scientific and Technical Potential of Coun-
tries of Africa (provisional edition), Field Science Office, Nairobi,
1969; UNESCO, Survey of the Scientific and Technical Potential of
African Countries, Paris, 1971.

Although the surveys are preliminary and incomplete, they are
instructive. The research institutions are mostly concentrated in a
handful of countries, but 19 of the 34 countries participating in the
survey had in existence government bodies or institutions responsible
for the formulation of national science policy. (See Table 3.7.)
These surveys also show an apparent misconstruction of national research institutions, like government and industrial laboratories and
universities, as panacea for primarily pure, and typically prestige
research. Elemental applied research that, together with pure research, constitutes the essential foundation of long-term technological development, has been a secondary and infrequent activity. This
practice represents a failure to recognize the limited roles of NRIs,

TABLE 3.7

African Countries with Government Bodies Responsible for Science
Policy Making, Planning, and Coordination, 1970

Ministry of Science[a] or National Council or Committee for Science Policy	Overall Science Planning Body	Multisectoral Science Research Coordination Body	Coordination Bodies for:		
			Medical Research	Agricultural Research	Industrial Research
Egypt	Egypt	Egypt	Sudan	Sudan	Sudan
Ghana	Ghana[b]	Ghana[b]	Nigeria	Dahomey	Nigeria
Mali	Niger[b]	Mali		Ivory Coast	
Niger	Nigeria	Niger[b]		Mali	
Nigeria	Senegal[b]	Nigeria		Niger	
Cameroon	Zaire[b]	Cameroon		Upper Volta	
Zaire	Ethiopia[b]	CAR		Zaire	
Congo (PR)	Tanzania[b]	Zaire[b]		Ethiopia	
Madagascar	Uganda[b]	Congo (PR)		Malawi	
Uganda	Zambia[b]	Ethiopia[b]			
Zambia		Madagascar			
		Tanzania[b]			
		Uganda[b]			
		Zambia[b]			

[a]Having no other responsibilities.
[b]The same body performs both functions.

Source: UNESCO, World Directory of National Science Policy-Making Bodies 4 (October 1970); and country reports.

particularly, where the coordination of their activities with national development plan is lacking. The relative effectiveness of NRIs is dependent on the extent and the manner in which their activities are defined by the objectives of the national development plan and on the integration of their implementation with that of the plan.

The lack of problem orientation in the sudden proliferation of government research institutes during the 1960s has also left them far underutilized both by the public and the private sectors. The general experience with these institutes has been contrary to the practice in private industry, where research is problem-oriented and dependent on the estimated time of payoff for the invested capital. However, in both private and public institutes, the associated risk factor is always present until the technological (product or process) innovation succeeds in the final product, or service market, or its general use and diffusion are amply demonstrated.

Major improvements in the effectiveness of NRIs also requires a link between "know-why" or science and technological know-how. In Chapter 6, Theodore W. Schlie presents one of the few thorough analyses of this link. In his exposition he examines this link as a "vertical" transfer of technology, namely the transition from fundamental research to applied research and on to R and D and engineering applications. A major factor in the establishment of this link is engineering design; namely, the application of a new scientific knowledge to meet needs through a new process or a new product or both. As the applied phase in the development of a new engineering technology, engineering design is vital to a better understanding of the technological requirements of production by private industry. On the whole, engineering design represents an important part of the scientific and technological infrastructure that is essential for the adaptation of imported as well as the development of indigenous technologies, the production of replacement parts, and the repairs and maintenance of machines and equipment.

Ancillary to engineering design activities is the machine tool industry. This industry and the engineering design capabilities in many African countries are relatively undeveloped. The few countries that have begun to demonstrate development of capabilities in both areas are Ghana, Guinea, Mali, Ivory Coast, and Cameroon. Figure 3.1 shows simplified, but basic stages of scientific and technological innovation activities. Many research institutes in African countries are currently locked into the basic research phase for all practical purposes of indigenous technology development. Although scientific knowledge is a requirement for advanced technological development, its limited application and nonmission orientation in NRIs has tended to limit the extent of its practical benefits to society. This experience and the generally restricted environment in which

FIGURE 3.1

Scientific and Technological Innovation Linkages

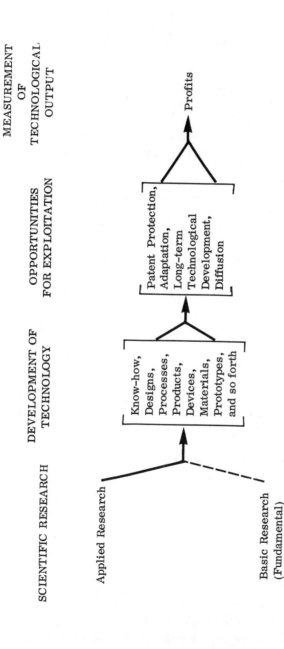

| SCIENTIFIC RESEARCH | DEVELOPMENT OF TECHNOLOGY | OPPORTUNITIES FOR EXPLOITATION | MEASUREMENT OF TECHNOLOGICAL OUTPUT |

*This measurement, which is based on commercial use, may be viewed not solely in terms of profits but more fundamentally in terms of general use and as reflected in growth in, for example, total factor productivity, per capita income growth.

Source: Adapted from J. B. Quinn and J. A. Mueller, "Transfering Research Results to Operations," Harvard Business Review 41 (January–February 1963): 59.

basic and applied research projects are conducted, including those
in universities, also suggest the need to devote major attention to the
sociocultural dimensions of indigenous and, more importantly, imported technology. The simultaneous consideration of the hard and
the soft factors, namely, the engineering and the sociocultural factors are particularly essential in determining the feasibility and the
desirability of imported technology.

As indicated earlier in this chapter, a dominant volume of the
expenditure on scientific research in virtually all African countries
is undertaken, directly or indirectly, by government agencies. Under
the principle that scientific research that cannot be developed into
technology, process or product, constitutes a commercial failure,
research activities in DCs are characterized by high risks, a high
rate of failure, and high costs. Because of such costs consideration,
the required duration for scientific and technological manpower
training, and the associated problems of personnel attrition, foreign
companies are somewhat reluctant to undertake major investment in
the development of scientific and technological manpower. Besides,
this manpower is considered part of the basic industrial infrastructure that these companies expect to be developed through national
efforts. The common practice of staffing key positions in the subsidiaries of foreign companies in the DCs by expatriates instead of
indigenous personnel has been defended by the short supply of indigenous scientific, technological, and entrepreneurial manpower in
these countries.

In the absence of systematic forecasting of future technological
needs, the problem-solving roles of research institutes primarily in
the areas of adaptive research on the immediate absorption of imported technology and the development of indigenous technology anew
are likely to be undirected. In the latter, a general lack of direction
is likely to contribute more to the typically high risks and expense
associated with the development of new technology.

The sketchy evidence on performances in specific research
activities--for example, food processing and chemicals--in a sample
of countries, namely, Nigeria, Ivory Coast, Ghana, and Senegal,
suggests that the technological output per unit of financial and scientific input expended are generally low when compared with similar
industries in other DCs, for example, in Latin America. Based on
the limited empirical evidence in the Nigerian case, the effects of
research and development on the national economy also indicate that
the output of research has been less than commensurate with the consumption of resources in research activities.[4]

The practical application of scientific knowledge and the systematic use of that knowledge for the production of materials, methods,
and processes, and the necessary engineering development of an

operational production system, all of which constitute the basis for R and D, are still rudimentary in both the private and the public sectors. The minimal local R and D activity that typifies most African countries and the limited opportunity for the acquisition of experience in this area also reflect the common practice whereby the engineering configurations and associated processes of imported technology are predetermined primarily by the fundamental requirements in the exporting countries. A further constraint on the maximization of benefits from investment in research in the restricted adoption and diffusion of new technological know-how. This, in turn, may be attributed to the lack of appropriate entrepreneurial and organizational links between NRIs and private industries. The remoteness of research from the direct needs of industry is inherent in the isolation of NRIs, usually in universities, from the practical problems of society. However, there are a number of exceptions. One apposite example is the Rubber Research Institute of Nigeria, which after a number of years has become actively involved in bringing about major improvements in the rubber industry through production and distribution of high yield disease-resistant plants to rubber farmers at subsidized prices. One of its other diffusion roles is the dissemination of information on improved cultivation methods.[5]

The broad perspective that emerges from the foregoing is that there is a general lack of synergy between NRIs and industry. The evidence presented illustrates the relatively undeveloped channels through which the technological needs of the growing number of industries in African countries could be fulfilled and shows the serious shortcomings of the existing weak linkages for any advantageous use of technological know-how. The establishment of the pertinent links requires a number of provisions: the development of scientific and technological manpower attuned to national needs in private industry and public enterprises; consideration of the significance of scientific and technological developments on the basis of both their social utility and economic profitability; the mobilization of appropriate scientific, technological, entrepreneurial, and managerial manpower in concert with changes in national science and technology policy; and the establishment of trust by private industry in research institutes.

In the absence of established trust and confidence by industry, the roles of NRIs, no matter how well designed, are significantly constrained and opportunities for contractual technological services as well as the smooth diffusion of technological innovations are likely to be impeded accordingly. Since in many cases NRIs are government or quasi-government establishments, they are subject to the prevailing budgetary conditions and political directives. These conditions tend to be counter productive for collaborative research between the public and the private sectors. Although the continuing support and in many cases the supervision of NRIs by government ministries or councils are inevitable under the current conditions, a major improvement in the minor role played heretofore by private industry in the development of scientific and technological manpower is equally exigent for optimum collaboration.

The extent to which the critical size for the optimum utilization of NRIs, and for viable scientific and technological development, are attainable in many of the landlocked African countries, is difficult to ascertain. More importantly, policies on a national piecemeal basis involving some of these countries have promoted unnecessary duplication of efforts and inefficiencies in the utilization of scarce financial and human resources.

Although many African countries have demonstrated political viability in spite of the balkanization of the continent, a good number of the small landlocked countries like Mali, Upper Volta, Niger, Chad, the Central African Republic, Rwanda, Burundi, Malawi, Botswana, Lesotho, and Swaziland are threatened by economic inviability. But some new possibilities are being discovered in a few of these countries. For example, there are possibilities for oil in Chad and rich mineral deposits in Botswana. Nonetheless, their being landlocked requires major consideration. This condition and their relative poverty have often been identified to underscore the merits of formalized regional economic cooperation among African countries. A critical component of such cooperation will inevitably be in scientific and technological development.

CONCLUSION

The occurrences of stringent foreign investment policies in African countries have been few compared to existing codes in other DCs in Latin America and Southeast Asia. However, the considerable dominance and control exercised by international companies in Africa are likely to frustrate efforts at setting stringent investment codes on a broad scale, at least in the next decade.

The success of foreign investment policy, measured by increases in the direct participation of the domestic population in industry and commerce, and by the social utility of this participation, is dependent on the integration of foreign investment policy with science and technology policy and national development planning. The attainment of the necessary integration requires numerous institutional factors, some of which have been examined above in this chapter. Included among the fundamental instruments in the establishment of this integration are NRIs. The effectiveness of NRIs including universities is greatly enhanced by their functional link with industry. Given the importance of the engineering phase in the development of indigenous technology and the adaptation of imported ones, the need for division of labor between universities and NRIs on the one hand, and among NRIs on the other is extremely essential in, and among, African countries both in terms of minimizing unnecessary duplication of efforts, and promoting collaboration and efficiency. Although this may be fraught with major problems of policy formulation and coordination, in addition to the typical scale problems, they may be resolved by taking advantage of the merits of such division of labor including the opportunity to maximize the utilization of existing scientific and technological manpower and research facilities, increase communication among the national and regional scientific communities, and facilitate the diffusion of new knowledge.

Ordinarily, the concentration of scientific training and pure research in universities and colleges of science and technology, and the concentration of applied research and development in research institutes and in the future in private research laboratories merit special coordination. The implied institutional division of research activities requires a rapidly expanding scientific community with the capability to influence government policy and planning.

One additional requirement for improvement in the limited communication among NRIs is the development of appropriate institutions to serve as the hub for national or regional documentation on S and T matters that bears on the special problems of African countries. This is vital for the efficient diffusion of scientific and technological information.

NOTES

1. United Nations Educational, Scientific and Cultural Organization, International Conference on the Organization of Research and Training in Africa in Relation to the Study, Conservation and Utilization of National Resources, Lagos, Nigeria, July 28-August 6, 1964. See also Outline for a Plan for Scientific Research and Training in Africa, UNESCO in association with UN-ECA, Paris 7e, 1964. The Yaounde Symposium on the Promotion of Scientific Activity in Tropical Africa, Yaounde, Cameroon, 1967 UNESCO (NC/SPS/11, SC/SP.68. XII.II/a), Paris 7e, 1969.

2. UNESCO, Scientific Research in Africa: National Policies and Research Institutions, UNESCO, Paris 7e, 1966; and Enquete Sur le Potential Scientifique et Technique des Pays d'Afrique, UNESCO, Paris 7e, 1971. See also United Nations, Survey of Economic Conditions in Africa, 1971, Part I U.N.-Economic Commission for Africa. Addis Ababa. (New York: United Nations, 1972), p. 43.

3. United Nations, Survey of Economic Conditions in Africa, 1971, p. 48.

4. The empirical basis of this conclusion can be found in D. Babatunde Thomas' Capital Accumulation and Technology Transfer: A Comparative Analysis of Nigerian Manufacturing Industries (New York: Praeger, 1975), chapter 3. Typically, a significant proportion of allocation for research, in some cases as much as 90 percent, go to pay for staffing and only a minor part of the resources go into acquisition of appropriate equipment for functional research facilities. Evidently, an optimum combination of staffing and physical research facility is most desirable for productive research activities.

5. African Development (African Buyer and Trader Publications Ltd., April 1975), p. 71.

4

MULTINATIONAL
CORPORATIONS' ROLE
IN THE CHOICE OF
PRODUCTS, CHOICE OF
TECHNIQUES, AND
THE SUPPLY OF
TECHNOLOGICAL INNOVATIONS

INTRODUCTION

In the absence of a consensus on what constitutes a multinational enterprise/corporation, or MNC, the conceptual scope in this
chapter includes the consideration of all foreign companies operating
in DCs, but the primary focus will be on MNCs based on the following
definition.

For the present subject of inquiry, a MNC is any parent company, with majority ownership by nationals of one country, that controls a large cluster of corporations, affiliates/subsidiaries, of
various nationalities.[1] Such a company is an economic entity usually
with a global organizational strategy. In addition to MNCs, transnational corporations and the residual group of foreign companies
operating in DCs are considered to constitute the totality of the channels through which proprietary technology is supplied and transferred
from the ACs to DCs. The transnational corporations operating in
Africa, namely Shell-BP and Unilever, are for all practical purposes
MNCs in their orientation.

By their nature, and as evidenced in their business operations,
foreign companies operating in African countries tend to be home-
country, globally, or oriented strictly toward the parent company
instead of the host country, and basic information for a thorough
assessment of their activities in these countries is rarely accessible.[2] Consequently, consideration of these companies' role in the
transfer-acquisition of technological innovations to African countries
has tended to be more notional than empirical. This has inevitably
resulted in a great deal of controversy, some of which will be examined below.

Of the various groups of foreign companies, MNCs constitute
the single most significant channel through which products, capital

investment--portfolio and DFI in plant and equipment, technological
and organizational know-how move between and among countries.
Although the number of MNCs with direct business interests on the
African continent is relatively small when compared to the number
of all the other foreign companies, the size of their business operations in terms of sales and their overall investment is much more
substantial when all their activities through "third party" firms are
considered. All other foreign companies do not constitute corporate
clusters as do MNCs. These companies tend to be smaller than the
subsidiaries of MNCs and their ownerships vary considerably; including European, Indian, Syrian, and Lebanese origin.

Apart from differences on the basis of the sectors in which
foreign companies operate in DCs, principally extractive, services,
and manufacturing, fundamental differences also exist among these
companies in the manner of their search for economic opportunities.
The MNCs tend to seek and establish production outlets and markets
that fit into some overall global strategies, while other foreign
companies ordinarily have no similar explicit global strategies.
Nonetheless, when the latter groups of companies are in sufficient
numbers in individual host countries, for example, the large numbers of Indian, Syrian, and Lebanese firms in east and west African
countries before the early part of the 1970s, and each has the optimum firm size for a market or industry, they are capable of constituting, as à group, a viable competition that local firms are still
not capable of mustering or sustaining against MNCs. In the more
prevalent cases, where foreign companies are few in number and
competition is weak or nonexistent, advantages of earning monopoly
rents on the basis of past technological choices by the few foreign
firms are buttressed, and opportunities for active participation by
local firms are further constrained. Generally, in African countries, and more so when compared to Asia and Latin America, foreign companies enjoy built-in protection against local competition
because of the relatively small size and number of indigenous firms
and the rudimentary stage of modern entrepreneurial development.

In the three aspects of MNCs' role to be examined in this
chapter, namely the choice of products, the choice of techniques,
and the supply of technological innovations to host countries, most
of the analyses also apply to other foreign companies. One characteristic they all have in common is the limited extent of their host-
country orientation. This is a paramount source of conflict between
the DCs and MNCs. It is fundamental to their asymmetrical perspectives on the development consideration of the social utility,
technological significance, and economic profitability of DFI in these
countries.

Although the manner of decisions on choice of techniques,
choice of products and output mix are clearly elemental to this

conflict, a more fundamental consideration is that on the one hand, DCs are seeking the means to implement their long-term national development plan objectives, and on the other, MNCs have as their primary goal short-run maximization of global sales and profits and the minimization of tax liability and other costs. Among the approaches used by MNCs to attain these objectives include the diversification of their avenues of operations to minimize risks, and the use of transfer prices. The latter may entail the charging of their affiliates that exist in high-tax countries corresponding high prices for goods and services supplied to show low profits, and those in low-tax countries low prices for identical goods and services to show high profits. (Of course, it would be useful to find out what is stopping large numbers of MNCs from setting up dummy companies in low-tax areas at least as sales outlets in order to take advantage of selling at competitive prices and earning high profits.)

Of course, many MNCs are not advantageously placed to have affiliates in both low- and high-tax countries and thereby maximize the benefits of variable transfer pricing. Two basic arguments that seek to explain the existing relationship between the DCs and MNCs have the obvious coloration of the asymmetry of their goals. Viewed from the standpoint of the DCs, it is the general lack of willingness on the part of MNCs to eschew the goal of short-run profit maximization that has tended to exacerbate the conflict and this has led to a wave of governmental nationalization of DFI in the interest of long-term national development objectives of the host-countries. From the MNCs' viewpoint, the existence of political instability and the possibility of nationalization entail high risks that require the maximization of short-run profits wherever possible to ensure the "proper" return on the invested capital. Apart from data problems, the merit of any empirical verification of the way both arguments fare is beset by "the chicken or the egg" type of relationship. Despite these ambiguities, one important observation has yet to be controverted fully; namely, that the MNCs' financial, organizational, and technological capabilities, their relative insulation from price competition, and overall market power, equivocal public policies by host governments, and the use of foreign policy as a leverage usually to its full measure by the MNCs' home government have all ensured the MNCs' ability to exercise considerable independence in their operations in a number of African companies. There are cases in African countries where most of these factors have tended to work to the serious disadvantage of the domestic economy. The manner of MNCs' independence has yet to be documented by sufficient case studies of African countries. One example of major importance is Societe Generale in Zaire, formerly the Belgian Congo. Considerable amount of independence was exercised through Union Miniere in production and pricing decisions on copper. In response to its exclusion from such major decisions, the then Zairian government moved to nationalize Union Miniere, but this effort failed because of the shortage of necessary skilled manpower to replace expatriates and more importantly, Societe Generale, the holding company for Union Miniere partially, but successfully, blocked the direct sales of Zairian minerals in the world market.*

The general trend in current thinking of African public policy makers is that the major sources of conflicts between the MNCs and the host government may be eliminated gradually through the promotion of joint ventures between public enterprise and MNCs. However, there is evidence suggesting that MNCs are generally not ready to take such a collaborative posture on a major scale. One dominant argument by many MNCs is that for efficient global planning, complete control over their subsidiaries is compulsory. This entails, among other things, the centralization of decisions on the choice of products and the choice of techniques. Two important realities of this weak disposition toward joint ventures are the recognition that along with joint ventures comes, by definition, manpower participation and the direct involvement of local personnel in the management and organizational decision-making process. The latter poses a major problem to the MNCs' secrecy on technological know-how and management habits that may not be in consonance with the host country's national interest. Furthermore, the ability to sustain their competitive edge in the market served by the subsidiary is viewed in terms of limited mobility of the local personnel to newly developing but competing enterprises.

THE CHOICE OF PRODUCT AND OUTPUT MIX

The collectivity of foreign companies in DCs clearly constitutes a powerful economic and political force. In addition to supplying or transferring technological innovations, new techniques and processes, principally through DFI and licensing agreements, these companies also transfer and dictate tastes for consumer goods. But, the constituent know-how in the new products and processes is not necessarily subject to automatic diffusion. At best, these transfers have tended to bring about superficial changes in the social and economic structure of the societies affected, and the impact of the technological innovations have been most evident on final consumer products.

---

*Note, however, that Societe Generale does not fit our earlier definition of MNCs and Union Miniere was operating strictly in the extractive industries.

A common theme in various studies seeking to explain why European and U.S. MNCs have not developed in Africa and contributed to fundamental or development-oriented structural changes at least to the same extent they have in South America and Asia is threefold: the relatively slow rate of requisite infrastructural development including basic institutions, such as financial and legal; the small market sizes in the national economies; and the asymmetry between the dominantly profit-maximizing objectives of MNCs, and the host governments' goals of expansion in domestic employment, value-added, and income growth.[3] On the basis of the development experience in the last two decades, there is ample evidence suggesting that the various transfers of technological innovations, particularly to African countries, have not been motivated primarily by the real development needs of these countries but by the MNCs' search for new resources and new export markets.

The marketing approach of the MNC is usually sophisticated and standardized globally or regionally, so numerous markets are served also by standardized products that are predetermined in the home-country, and the incremental cost of serving additional markets is likely to be minimal, notwithstanding the stage of the product life cycle in the initial market(s). The network of global markets provides MNCs with the opportunity to take full advantage of economies of scale. Furthermore, through their sophisticated marketing approach, MNCs have successfully stimulated consumer demand in the DCs, and thus lengthen the life cycle of products designed and developed basically for markets in the ACs. The adoption of products and techniques of production by DCs are considerably influenced by the regional marketing strategies and the "global reach" of the MNC, more so than by these countries' national development objectives. Thus, the seeming uncertainty about the market demand for the MNCs' standardized products are resolved, and the risk factor associated with producing locally rather than exporting or licensing the technological know-how to another company is markedly minimized.

Theoretically, the ability to penetrate foreign markets is dependent on the recognition of cultural differences and the need for varying products to suit individual markets. Nonetheless, national markets in African countries in durable and nondurable consumer goods have been penetrated by the standardized products of foreign firms through skillful marketing. The array of products is invariably more suitable for the affluent segment than for the masses in these countries and illustrates the "corrupting ethos of progress."[4]

Although decisions on the choice of techniques and the profit purpose are usually evident, the social purposes of the choice of products are not so evident. From the MNC's standpoint, the choice

problem is a major consideration if the company is vertically integrated with a large number of product lines that cannot be efficiently
produced in their entirety by each of the subsidiaries or affiliates.
Therefore, specialization and, wherever possible, economies of
scale are required among its affiliates. Consideration for proper
product mix is thus based on the MNC's global strategy to serve its
worldwide markets. In such instances, short-run cost considerations tend to override the need for the adaptation of standardized
production techniques. The manner in which consumers' demand in
DCs is stimulated for the products of industrial societies imposes
inappropriate choice of products and choice of techniques, and distorts other choices--for example, choices related to the pattern of
industrial development that are fundamental to national development.

CHOICE OF TECHNIQUES

Already delineated earlier are some of the advantages MNCs
have over other international companies and more significantly over
local firms. Next, we examine a major component of these advantages that by most measures is the primary explanatory variable for
the superior performance of MNCs. In its composite form, it consists of advanced organizational, managerial, and technological
know-how. These three factors are interconnected and they constitute the basis of technomanagerial decisions on the choice of techniques. Despite the superior performances of MNCs, the experience
of host governments and the results from their search for the attainment of fundamental national development objectives, such as infrastructure development, rapid employment expansion, income growth,
and long-term technological development are generally dismal. The
technological and managerial considerations in the decisions by
MNCs tend to diverge significantly from the purely socioeconomic
and political considerations in the desires of the host countries. One
of the results of this divergence is the choice of inappropriate technology by the former, given the domestic input structure in the DCs.
The existence of an asymmetry between MNCs and host governments in the pursuits of their respective goals has been recognized
above as fundamental to the change in public policy in DCs toward
foreign investment. To understand why these differences exist, and
what are some of the other basic solutions to them, it is important
to examine the forces at work in the choice of techniques by subsidiaries and affiliates of MNCs in DCs. In many cases in Africa, decisions on technological choices are dictated primarily by economies
of scale and other physical requirements external to the host country.
Among the other factors likely to influence the choice of technology

favorably from the standpoint of the DCs are the price elasticity of
demand for the firm's product and the elasticity of factor substitution. The extent to which the elasticity of factor substitution influences these choices is different from industry to industry. There
are some industries, primarily light manufacturing and ancillary
service areas like product handling and packaging in which factor
combinations could be made to reflect their relative availabilities in
the host countries through minor adaptation of production techniques
and at a minimal incremental cost. The contribution of the factor
combination to employment expansion in such cases is likely to ful-
fill some basic social utility as well as ensure economic profitability.
A reference industry in this respect is the pharmaceutical industry.
There are other industries, principally heavy manufacturing, in
which capital intensity is inherent to an entire plant or subsets of
operations because of the nature of known efficient production techniques, and scaling down that is usually a requirement for labor-
intensive techniques may not be feasible, though socially desirable.
Theoretically the prevailing technology and its total factor intensity
would depend on the relative factor availability and the scope of the
manufacturing operations.* Since in DCs the scope of operations
tends to be partial by being limited or concentrated in the final phase
of the manufacturing process, for example, assembly operations,
packaging, and in general the combination of intermediate inputs, a
production technique designed for the entire scope of the operations
and that reflects relative factor availability in the ACs is a source
of distortion in factor combination given the differences in factor
availabilities in the DCs and the ACs. Although technological choices
are influenced by relative prices of inputs in the theoretical sense
and by the size and growth potential of the overall global market,
they are more significantly influenced by the "state-of-the-art" or
the stock of production technologies that have been proven convenient

---

*It is conceivable that processes within a plant will vary; some
capital intensive, others labor intensive. The usual question is how
much can the latter stages of operations be expanded and the former
stages reduced to the barest minimum in terms of the total factor
intensity given the domestic input structure. The distinction, or
dichotomy, between labor and capital intensity should be advisably
viewed here, because of some of the confusion in the theoretical and
empirical analyses of various forms of input/output relationships.
For our purposes, technical choice is pluralistic, not dualistic.
When the choice of techniques is discussed in the latter sense, only
a particular stage in the production process receives attention--
usually the final stage.

or profitable in the experience of the ACs. These are not necessarily
suitable from the standpoint of high employment-oriented choices
dictated by the input structure in the DCs or by their long-term national development objectives. The tendency for the choice of product for the market in the DCs to precede the choice of techniques has
resulted in incompatibility of the techniques and domestic factor proportions. Usually, the techniques require large capital investment,
but result in minor change in direct employment. In most African
countries, wage employment typically associated with sectors in
which foreign investors are actively involved has remained relatively
low throughout the 1960s and the early 1970s. See Table 4.1. The
generally low-employment multiplier in these sectors is also reflected in the generally high unemployment figures shown in Table
4.2. Some of the increase in the recorded unemployment figures is
due, in part, to improvements in statistical gathering, but it is
largely attributable to the limited capacity for employment creation
in the major sectors. These influences are evident in the MNCs'
preference for sophisticated, latest vintage, and usually capital-
using technology in most of the stages of plant production. The opportunity to spread the relatively high-fixed cost associated with
capital-intensive technology and the risk factor associated with the
search for new, but "competitive labor-intensive technology," with
which the MNC has little or no current familiarity, are major considerations in the choice process.* Distortions in input prices due
to imperfections in capital markets also facilitate decisions on the
choice of techniques that, in turn, distorts priorities for national
development planning. These conditions are the primary explanations
of why the transfer of technological innovations to DCs in general is
a necessary but insufficient condition for long-term technological
development. The lack of compatibility of the choice of imported
technology with local conditions and needs where defined by national
development objectives, underscores the necessity for adaptation.
In addition to production costs, the extent to which the degree of
price or brand name competition in the particular industry influences
decisions on the choice of technology is an empirical question that
can be verified only on the basis of case studies given the differences
in the extent of the shelter and incentives traditionally provided to
foreign companies in different countries.

More importantly, appropriate adaptation is necessary for acquisition and facile diffusion and long-term technological development.
The disinclination of MNCs to adapt their technology readily to the
input structure and other prevailing conditions in the DCs is commonly

---

*The concept of "competitive labor-intensive technology" is an
outgrowth of current research by the author.

TABLE 4.1

Wage Employment Related to Population in
Selected African Countries

Subregion and Country	Wage Employment as a Proportion of:		
	Wage Employment (in thousands)	Population (in percent)	Estimated Labor Force (in percent)
West Africa			
Dahomey (1967)	30	1.2	3.0
Gambia (1969)	16	4.5	11.2
Ivory Coast (1969)	270	5.5	13.8
Mauritania (1970)	18	1.5	3.7
Niger (1969)	70	1.8	4.5
Nigeria (1970)	1,385	2.9	7.2
Senegal (1968)	133	3.6	9.0
Togo (1970)	33	1.8	4.5
Central Africa			
Gabon (1969)	62	12.7	31.7
Zaire (1967)	1,035	5.3	13.3
Rwanda (1967)	85	2.5	6.2
East Africa			
Botswana (1967/68)	28*	4.7	11.7
Kenya (1970)	1,064	9.5	23.8
Lesotho (1969)	20*	2.2	5.5
Malawi (1970)	160	3.4	8.5
Mauritius (1970)	122	14.6	36.5
Swaziland (1969)	49*	12.0	30.0
Tanzania (1969)	368	2.8	7.0
Uganda (1970)	308	3.1	7.7
Zambia (1970)	372	9.0	22.5

*Excludes persons working in South Africa.

explained by the complexity of the imported capital-intensive technologies that are generally developed for a mass-consumption society. Given the scale problem, they tend to require high cost for the required adaptation. The dominance of such arguments, which are usually couched in terms of the importance of economies of scale, has tended to retard possible progress toward the development of alternatives under which the selective scaling down of imported technology could satisfy certain well-defined utility-efficiency criteria in DCs. An important question that arises is, what sorts of incentives are likely to encourage MNCs to adapt their know-how based on the prevailing structural input conditions in the DCs? Attempts to answer this question will be made in the next two sections.

TABLE 4.2

Recorded Unemployment in Selected African
Countries, 1960, 1965, and 1968-70
(in thousands)

Country	1960	1965	1968	1969	1970
Cameroon	0.39*	0.24	3.48	--	--
Egypt	288.0	--	244.8	--	--
Ghana	11.3	11.3	17.6	15.0	17.6
Madagascar	0.77	0.84	0.85	--	--
Malawi	--	1.20	1.74	1.25	--
Mali	0.12	0.13	0.40	0.47	0.26
Morocco	21.6	18.6	22.9	27.4	--
Mauritius	2.25	8.33	9.08	14.28	20.66
Nigeria	6.86	20.94	12.93	12.17	13.82
Sierra Leone	7.89	11.56	14.11	14.73	16.14
Zambia	2.70	17.56	12.91	15.31	10.64

*1961.

Note: Certain of the figures only cover a limited area such as a small number of towns in the particular country.

THE CHOICE OF TECHNIQUES AND THE ACQUISITION
AND DIFFUSION OF TECHNOLOGICAL
INNOVATIONS IN DCs

Ordinarily, several distinct possibilities are open to foreign companies in their search for market opportunities in DCs. These possibilities are time-phased in their relationship to the technology transfer process and usually are manifested in the decisions of foreign companies to export finished products to the DCs' markets, or license the pertinent technological know-how to another company that chooses to invest and produce the products in the DCs, or undertake the DFI and production in the DCs through a wholly owned or majority-owned subsidiary or affiliate. The first option has been on the decline because of export control and the limited opportunity it provides for the acquisition of technological know-how given the level of scientific and technological infrastructure in most DCs. Since the 1960s, the DCs have been insisting on local manufacturing, and from their standpoint, licensing may also be desirable in terms of cost because the fees can be prearranged. However, its desirability depends on the relative equality of the bargaining power in the negotiation of the licensing agreement. One possible disadvantage of this option is that it may not provide a complete package in the manner that DFI can in instances of joint ventures with a local firm, where the arrangement covers equity participation as well as managerial and technological manpower development. The third option tends to be the most commonly used in view of the opportunity it provides the foreign company for closer control of the technological know-how, as well as the organizational, managerial, production, and marketing operations. From the MNC's standpoint, DFI and production in foreign markets also provide the opportunity to exploit the capacity of owned as well as known and accessible technology far beyond the scope and the life of a purely export arrangement. But, this is rarely used to the fullest extent possibly due to the proprietary nature of virtually all commercial technology.

The network of affiliate/subsidiary to parent company information flow and vice versa--for example, information that helps constantly to improve knowledge about the general character of the markets being served, as well as identify the strengths and weaknesses of government policies in the host countries--is vital for the desired control by the parent company.

Since the 1960s, this option has also provided opportunity for the maintenance and expansion of foreign markets in response to the increasing export control in the DCs. Although in manufacturing, for example, primary consideration is given to the ease of access to the sources of inputs, in practice the final or intermediate product is ordinarily produced for export by the parent company or another foreign subsidiary/affiliate, where the production technology and the product are both considered superior. The incidence of this practice is prevalent in African countries when compared to Latin America and Southeast Asia. For example, in 1970, value added to GDP by manufacturing ranged from 1 to 20 percent among African countries. This is in comparison to a range of 12 to 34 percent in Latin America and 6 to 28 percent in Southeast Asia.[5] In most cases, among African countries, the manufacturing industry has been contributing under 10 percent in value-added to GDP. This condition is also reflected in the trend in unemployment illustrated in Table 4.2. The reverse is the common practice in agriculture, where infrequent minor processing takes place at the primary source. In some instances, value added resulting from the major processing of agricultural products and raw materials, for example, rubber and oil, are undertaken abroad and the processed, or final, products are exported back to the primary producing countries.

In response to the need for the adaptation of techniques and when need be the products as well, attention has been routinely drawn to the technological knowledge necessary for production processes and products that belong in a special class, for example, automotive engines and computers. Yet these are products that have only minimal direct effects on the lives of the masses in the DCs. The experience in India with the Cummins engine for diesel trucks in his study on India has been pointed out in numerous other studies to illustrate the complexity of modern technology and the possible seriousness of underestimating the problem associated with the importation and adaptation of technological innovations in DCs.[6] However, this example and similar ones deal with special cases. Technological innovations in construction and food processing are less complex and easily adapted to the local conditions in the DCs. Furthermore, examples are rarely drawn from the manufacture and assembly of electronic components that utilize relatively labor-intensive techniques.

Since the profitability of the type of technology suited to the conditions in African countries have yet to be demonstrated, incentives that provide for the minimization of risks associated with the development of such a technology need to be carefully examined and viable ones encouraged. Normally, in the DCs, the government is looked upon solely for the promotion of such incentives. However, proper workings of the market forces are also capable of providing the necessary incentives. For example, the demand for appropriate technology in agriculture has been expanding in the last decade and this has been recognized by Japanese suppliers of agricultural technology and equipment. Although experience from agriculture suggests that the limited extent of market may have been a major factor in the adaptation of imported technology to the appropriate scale, no similar trend has been observed in manufacturing. In the latter, partial evidence to the contrary is that legal factors limited market, and fundamental efficiency and competitive production criteria have tended to impede the adaption of imported technology. In the absence of effective competition and given the constraints on skilled factor availabilities, large-scale capital-intensive techniques in many stages of the production process are likely to be preferred to labor-intensive techniques in manufacturing.

Although the actual transfer of appropriate technology from the ACs to the DCs, and their acquisition, adaptation, and diffusion by the latter may alter the nature of the product cycle, it is potentially a source of net improvements in the balance of payments and in the unemployment problems of the former group of countries in the long run. There is evidence that MNCs have begun to recognize the opportunity that exists in the potentially vast market for more agricultural as well as manufacturing techniques of a labor intensive nature involving broader stages of production. The real shift toward adaptive technology development and specifically, competitive labor-intensive technology has taken a gradual but encouraging turn in the last decade as a result of the Japanese experience and the involvement of the Intermediate Technology group.*

In a study of Kenyan industries by Howard Pack, one important exception to popular opinion about the proclivity of MNCs to use up-to-date technology was observed in the adoption of labor-using, or, capital-saving, methods to take maximum advantage of low-cost labor.[7] Pack's study reported experiences in a number of manufacturing industries, including paint production, cotton textile, and food processing. One of his partial findings shows that some foreign companies do search for old equipment which is obsolete by production standards in the ACs but highly serviceable in DCs. More importantly, these companies do undertake the reconditioning and, presumably, the adaptation of the old machinery to take advantage of their labor intensity and low cost of production. The incidence of this practice in African countries is still the exception rather than the rule and tends to be industry-specific. The evidence in these few exceptions, also suggests that MNCs using the network of their affiliates are more likely to seek and employ potentially efficient old-model equipment or secondhand machinery more readily than the indigenous firms.†

----

*The ITGs' effort is primarily in searching for and developing intermediate but efficient labor-using technology to meet the needs of specific DCs.

†In some of Mira Wilkins' studies on U.S. MNCs, she has found that it was not uncommon for companies to export obsolete equipment to foreign markets in Latin America. The reason, however, was not to take advantage of labor intensity, in most cases, but rather to use equipment that was written off and would otherwise be scrapped in the United States. At one stage, old equipment was used on the assumption that the output was a less up-to-date product that was perfectly suitable for the protected, minimally competitive DCs market but not suitable for the U.S. market. Some of the products of this equipment include sewing machines, refrigerators, automobiles, and elevators. This practice was often discontinued because the required skills for repairs and reconditioning were not readily available, and the continuing use of obsolete equipment proved

Pressure for increasing labor absorption is clearly one of the possible outcomes from the new investment policies taking shape in many African countries, and the extent to which the apparent trade-off between employment expansion and total factor productivity is likely to be resolved is not likely to be solely through joint ventures between MNCs and public enterprise in the host countries, but through major collaboration for the development of efficient and competitive labor-intensive techniques that are also capable of employment expansion. Such techniques are by definition potentially superior in production when compared with other techniques using similar scale and any other technology. If any other alternative technique of production yields higher output using less inputs overall but similar scale, then the labor-intensive technique in use is inferior.

The dominance of expatriates in the critical phases of management in foreign firms and in key technical aspects of research and development or generally in the development of the know-why and the know-how in the manufacturing industry has tended to restrict opportunities for employment expansion and for "learning-by-doing" by indigenous personnel. This suggests that the preference for wholly owned DFI by MNCs is a manifestation of the desire for maximum control and, when necessary, for the preclusion of indigenous personnel where sensitive decisions are involved and the advantages of the MNCs could be eroded. Governments in DCs must insist on a reorientation of MNCs so due consideration is given to national development objectives in host countries, if the recent patterns of joint ventures between MNCs and public enterprise are to be mutually beneficial.

CONCLUSION

Existing relationships between DCs and MNCs and invariably the ACs have tended to promote assymetry of objectives and the dependence of the former on the latter two entities. The development of alternatives for symmetry of the host governments' objectives and those of the MNCs' requires a radical departure from current practices. In seeking to minimize the dependence of the DCs on the ACs, the economic development capabilities of the MNCs need to be thoroughly examined and harnessed for the economic development of

---

costly. This subject area is still a virgin territory as far as African countries are concerned, and a great deal of work needs to be done to evaluate the past history of this practice in light of present African experiences and their future implications.

host countries. Although the financial dimension of these capabilities tends to receive a great deal of attention in studies on the MNCs, the components of virtually all foreign companies' technological, managerial, and organizational know-how are equally important and represent a more complete picture.

A major precondition for the proposed departure and the optimum exploitation of these capabilities is a correction of existing disproportionate distribution of bargaining power which has continued to considerably favor MNCs.* One approach to such a correction is the regulation and stabilization of the bargaining process by means of multilateral policy actions guaranteeing similar quasi-countervailing power as recently evidenced in the cause of oil and bauxite.

For mutually beneficial relationships, the new order likely to emerge from the above proposed arrangement will require close collaboration between the host countries and MNCs in clearly defined phases of decision making by the latter on the choice of techniques, choice of products, and the transfer of technological innovations. Ancillary to this relationship is the promotion and maintenance of close collaboration between the governments of the host countries and the home governments of the MNCs in the use of fiscal measures or incentives for the following objectives: to ensure that host countries are equitably compensated for drains on their resources; to encourage the reinvestment of retained earnings in the countries where earned for plant expansions, which is in as much consonance as possible with both the host countries' national development priorities and the MNCs' prespecified objectives, and specifically to support national research institutes for R and D activities; and to maximize opportunities for participation by indigenous entrepreneurs. Nigeria and Ghana have begun to pursue the third objective, but progress on the others has been extremely slow. Clearly, the attainment of these objectives will depend on a better crystallization of the goals of the DCs in general than has been the case thus far.

---

*Although the development of a quasi-countervailing power in the case of oil and bauxite producer countries as compared with the MNCs has raised some doubts about the extent of this assertion, these recent developments and the natural resources involved represent a limited experience and could therefore still be considered the exception instead of the rule.

NOTES

1. The essentials of this definition are contained in Raymond Vernon, Sovereignty at Bay (New York: Basic Books, Inc., 1971).

2. For further discussion of the three distinctions, see V. Perlmutter, "The Tortuous Evolution of the Multinational Corporation," Columbia Journal of World Business (January–February 1969), pp. 9-18.

3. See Technology and Economics in International Development (Washington, D.C.: AID, May 1972) and United Nations, The Acquisition of Technology from Multinational Corporations by Developing Countries (New York: United Nations, 1974).

4. This characterization can be found in Theodore Roszak's introduction to E. F. Schumacker, Small Is Beautiful: Economics As If People Mattered (New York: Harper and Row, 1973), p. 7.

5. United Nations, Survey of Economic Conditions in Africa, 1971, part I, U.N.-Economic Commission for Africa, Addis Ababa (New York: United Nations, 1972), p. 125.

6. See Jack Baranson, "Transfer of Technical Knowledge by International Corporations to Developing Economies," American Economic Review (May 1966), pp. 260-61.

7. Howard Pack, "The Use of Labor Intensive Techniques in Kenyan Industry," in Technology and Economics in International Development (Washington, D.C.: Agency for International Development, May 1972).

CHAPTER

5

FOREIGN INVESTMENT
AND THE
ACQUISITION OF TECHNOLOGY:
KENYA AND
TANZANIA
Walter A. Chudson

In its broadest sense, the word technology refers to all forms
of knowledge that affect the productivity of the factors of production.
The world's supply of technology consists of two parts: knowledge in
the public domain, freely available to all who can use it, and privately
held or so-called "proprietary" technology, either protected by legal-
ly conferred monopoly through patents or trademarks or by secrecy,
and thus inaccessible to others except on terms acceptable to those
who possess it.

Most proprietary technology is held by private business enterprises, but some may be owned by state-owned enterprises in socialist or other countries. In the field of industry, particularly manufacturing and until recently extractive industry, a major method
through which technology is transmitted internationally, from one enterprise to another, is through expansion of the "parent" enterprise
in the form of foreign direct private investment--in its current manifestation, largely through the multinational corporation.

In this form, technology is supplied as part of a "package" consisting of: capital; technology in the engineering sense, product and
process technology; managerial know-how; capabilities arising from
marketing affiliations and various centralized facilities that may contribute to the efficiency of the total system in terms of its economies
of scale, generation of new technology through research and development.

The "price" or cost of technology supplied through direct investment is not explicit, even though a substantial part may be
classed as intrafirm charges in the form of royalties, dividends on
capitalized know-how, management fees and other accounting charges.
Under these conditions, after deducting a charge provided in advance
for the imputed cost of capital, the major part of the total profits

earned by the firm can be said to constitute a payment or rent for technology. The arbitrariness of these charges in intrafirm accounting is a well-known source of controversy just because, like the setting of "transfer prices" between two units of a firm, they affect the international allocation of profits and hence the tax receipts and foreign exchange transfers of the countries concerned.

When we turn to the objectives and policies of developing (host) countries in respect to the package of productive factors that might be imported through the multinational enterprise, or through alternative arrangements, we find ourselves confronted by a bewildering array of motives. This is not intended as a criticism. It is a warning against the pitfalls besetting discussions in which reasonably objective economic criteria, on which there are some grounds to expect a consensus, inevitably confront political, social, cultural, and philosophical value judgments. Any national development policy reflects the same confrontation. But the appraisal of the choice, development, and cost of imported technology, specifically through foreign business participation, adds another dimension. Questions of foreign control, technological dependence, different motivations of private suppliers and the host country, and fear of foreign dominance loom large.

In recent years, it has become clearer that the special characteristic of DFI is the technological component of the package. This does not imply that capital is not in short supply in developing countries. It does suggest that in a wide range of activities, particularly manufacturing, the problem of obtaining know-how is more difficult to solve than that of borrowing or mobilizing capital locally. There are few developing countries that are unable to mobilize locally or borrow ahead much of the capital required to construct and operate a viable factory, excluding massive mineral production and processing establishments. But the alternatives of mobilizing the human capital to perform the functions of production, marketing, training, and research frequently present barriers that can be surmounted only by using foreign carriers of technology in some form. Thus, acquisition of technology becomes a question of comparing direct investment through the multinational corporation with the alternative channels of foreign business participation.

Developing countries are not alone in their high degree of dependence on foreign technology, as exemplified by areas like Canada, Australia, and New Zealand, whose economic progress has depended heavily on foreign technology. This is either embodied in imported machinery, supplied through direct foreign investment, through licensing agreements, or other contractual arrangements between local firms and foreign owners of technology.

This is the circumstance that gives rise to a wide range of policy issues. In this chapter, I discuss some of the aspects that appear particularly relevant to the recent situation of Kenya and Tanzania.

The issues confronting a country, involving the use of direct foreign investment as a channel for acquiring technology, can be posed as follows: Assuming that the project is expected to yield a given net social return to the country, according to some measure of social cost/benefit analysis, is a private foreign investment the most appropriate channel in terms of the net return involved? Does it offer the most appropriate form of technology for the country's conditions and objectives? In the practical measurements of social costs and benefits, there are general problems of evaluation, assessment of external effects both benign and negative, and value judgments.

The assessment of external factors includes two types of benefits and costs: those that can be measured with reasonable objectivity in economic terms, including the opportunity cost of inputs and various distinct externalities of the project and those that reflect value judgments reflecting goals like political, social, and cultural autonomy, personal income distribution, and regional location. Even so, one is faced with the nagging question of what might have been the position if reliance on domestic resources were greater. What is the probable interaction between foreign inputs and the supply and productivity of domestic resources in the form of human skills, including entrepreneurial capabilities, domestic savings, research and development of appropriate technology. Is there a simple increment, a displacement, or a stimulus to the long-term growth of local resources by the injection of foreign inputs?

PROJECT APPRAISAL AND FOREIGN INVESTMENT POLICY

Any appraisal of a direct foreign investment includes an appraisal of the advisability of acquiring foreign technology. One can compare this with alternatives: varying degrees of equity participation, joint ventures; licensing; management contracts; technical service agreements; "turnkey" projects. We can say that formulating a policy toward DFI requires a framework for appraising all types of projected investments, domestic or foreign.

Under conditions in which prices of inputs and outputs reflect their correct social opportunity cost, the calculation of private rate of return on investment, say, through calculated discounted cash flow, is a satisfactory measure. Since this condition cannot be

assumed to prevail, governmental authorities are increasingly turning to systematic project appraisal, as a means of combining into a
single calculation various economic criteria indicating projects
capable of increasing output as much as possible using abundant resources, such as labor, intensively and economizing on scarce re-
sources--capital and foreign exchange. In making this analysis, account clearly should be taken of alternative technologies. If technologies or capital are not available locally, then the calculation should
include their foreign cost compared with their local cost of supplying
the equivalent, which may be infinite if the required technology is not
available.

Such an appraisal involves adjustment of market prices to reflect the impact of alternative uses of given resources on the economy
as a whole. The opportunity cost is thus measured as the value of
the benefit foregone by diverting resources from one use to another.
Then calculations also should include adjustment for costs or benefits that are external to the project, for example, pollution or availability of commodities at lower costs than before.

The relevance of this reasoning of the situation in countries
like Kenya and Tanzania is that there are solid grounds for assuming
that market prices diverge substantially from the opportunity cost of
capital, skilled and unskilled labor, foreign exchange, and various
local inputs. The implication is that the contribution of specific foreign investment projects in Kenya and Tanzania, as measured by
private rates of return, may diverge from the social profitability of
the project.

KENYA

The most obvious situation results from commercial policies
that support an overvaluation of the rate of exchange of the country's
currency. This provides a subsidy to firms manufacturing for the
domestic market that must be paid by the rest of the economy. In
Kenya, it provides a subsidy from which foreign firms benefit because they account for the major part of industrial production. This
benefit also is reaped by the local capitalist and wage earner in manufacturing, all at the expense of the agricultural producers. The
profits earned in manufacturing are increased by providing imported
inputs at low cost through overvaluation of the foreign exchange rate
and through duty-free importation while sales prices are protected
behind effective tariffs and import restrictions of extremely high
levels. To the extent profits are repatriated, the loss to national
income is greater than if the same "sick," high-cost industries were
locally owned.

This circumstance also tends to make imported capital equipment cheap in terms of world prices and, hence, to reduce the incentive to modify technology to take advantage of low labor costs. Wage levels are higher than they would be. If wages were to reflect labor's true social cost, they would be lower as discussed below.

There is ground to believe that much criticism of foreign investment in Kenya in recent years is based on the consequences of the commercial policy described. The villain of the piece is the high cost of import substitutes in the manufacturing sector.

In Kenya, public awareness and recognition of this situation has grown recently. In the Budget Speech for 1973/74, the minister for finance strongly expressed the intention of the government to reverse gears on tax and tariff policy to encourage efficient use of resources. The goal is to reduce and hopefully eliminate the highly protected manufacturing industry that has been built behind high tariff walls. This includes foreign exchange costs of projects higher than foreign exchange benefits.

The policy proposed would establish tariffs, or equivalent restrictions, that would be uniform for both finished manufactures and machinery, and imported intermediate inputs. Thus, allocation of resources to the latter industries would not be prejudiced. They now are by the combination of commercial policy and by special incentives through tax-free imports. On the export side, one must secure a balanced allocation of resources to that sector and "break out of the trap" of dependence on the growth of the domestic or the east African market. Thus, there will be instituted a system of export compensation payments to offset the cost disadvantage to the export sector, arising from the remaining import duties.

In terms of social cost-benefit analysis, the proposed reform of Kenya's commercial policy amounts to the establishment of a market valuation of foreign exchange that will be considerably closer to the present "shadow rate" based on the relationship of world and local Kenyan prices. This means a substantial step has been taken to bring the market profitability and social profitability of investments, including foreign investment, closer together. An important consequence of this would be that the need for special project appraisal techniques to "screen" foreign investments as such, in terms of their social-cost-benefit, will be greatly reduced. Market prices would give a reasonably close indication of the social-cost-benefit balance.

TANZANIA

No official expressions of concern about Tanzania's industrial policy, comparable to those of Kenya, have been expressed. Within

Tanzania's less market-oriented economy there is a necessity to adjust factor costs and the exchange rate so as to calculate social-cost-benefit of projects in order to economize scarce resources. This is recognized officially within the industrial strategy of the development plan. In the context of Tanzanian conditions, this means appraisal of projects using: shadow "prices" for foreign exchange, capital, unskilled and skilled labor, and selected domestic inputs. This method is more efficient than pragmatic tests often applied by government to foreign and domestic investments, such as their net contribution to foreign exchange, employment, and domestic value added. The reason is that it incorporates into single measure, internal rate of return, all economic goals that bear directly on resource efficiency. This permits a direct comparison of a given project with all others.

Using even roughly accurate shadow prices, this method should ensure rejection of all projects that do not give a return to national resources at least equal to the opportunity cost of capital. If the appraisal is carried out, Tanzania will also be weighing the economic efficiency of alternative technologies, taking into account not only alternative technological processes, but the relative real cost of labor and capital too.

THE USE OF LABOR-INTENSIVE TECHNIQUES

The question whether foreign investors or foreign technology suppliers provide "appropriate" technology comprises several aspects. The most "classical" is the factor-proportions question: do foreign technology suppliers adapt technology to the factor-proportions of the developing host country? Following are a few comments on recent evidence in relation to Kenya.

Case studies in Kenya comparing foreign enterprises with local firms, in sectors where there are both kinds of company, indicate, contrary to expectations, that foreign enterprises seem to be less capital-intensive. During the same time, foreign enterprises had lower labor cost per employee. Speculating, foreign enterprises have more skilled supervisory staff, which would allow them to use production techniques utilizing more low-cost, unskilled labor.

Another example of comparing the cost of tin can production revealed that costs were higher for a more labor-intensive, semiautomated method than a fully automated process. This comparison was based on market prices for factors of production. The study indicated the "shadow" price of labor must be very low and subsidy very high to make the labor-intensive technique profitable.

These sketchy examples fail to do justice to problems of technological choice in relation to real costs of labor/capital. One must take into account the motivation and possibilities of the business firm, whether foreign or domestic.

There are some reasons why the adjustment of factor prices to reflect domestic opportunity costs might not suffice to stimulate substitution of labor for capital in manufacturing production. One is that in the absence of competition, that is, Kenya and Tanzania, there is not a strong incentive to adjust processes to reduce costs. This may be outweighed by reluctance to experiment with new methods. The recently projected reform of commercial policy in Kenya should induce a heightened competitive climate. To the large multinational firm, the technology previously developed in its home country represents a sunk cost. Hence, anticipated gains from developing and adopting a more labor-intensive technology will prove substantial in order to induce time and resource investment in significant product process modification. Significant innovation by private firms has been largely in response to the promise of a large market. This is not likely to be the case in developing countries unless an export-oriented industry is involved, or a product sold throughout the world.

Second-hand equipment is a potential source of relatively labor-intensive technology. However, governments of many developing countries tend to discourage its importation, especially by local enterprises.

However, foreign firms are better equipped to evaluate and buy such machinery.

THE PRODUCT MIX

In macroeconomic terms the kind of product produced is likely to have far more influence on the proportion of labor and capital employed in the economy generally and in industry in particular. Some industries are inherently labor-intensive. Others are inherently capital-intensive, although some choice of technique may be available.

Since expanded employment is a major objective in Kenya and Tanzania, this aspect of the question is important.

Perhaps the most significant contribution to a shift in the product mix would be the change in commercial policy in Kenya to which reference was made earlier. This would tend to reduce a number of import-substituting industries that are relatively capital-intensive, while at the same time encouraging export industries, including processing of agricultural products, which are relatively labor-intensive. Similar policies, or their equivalent, in Tanzania would presumably have the same effect.

At the present stage of development in Kenya and Tanzania, it is difficult to foresee clearly the early emergence of the type of manufacturing for export that has been such a striking phenomenon in a number of more advanced developing countries in recent years. It is in these sectors that the interests of the multinational enterprise and the host country most overlap. Both parties are interested in employing labor, and the multinational enterprise offers the critically scarce elements of know-how in the form of product design and marketing contacts in export markets. But for this kind of production to be competitive, a highly disciplined labor force is required, not merely low wage rates.

While the growth of exports of manufactures from countries like South Korea, Taiwan, Singapore, Hong Kong, Brazil, and Mexico has been hailed as a break away from inward-looking industrialization, some features of this development have been viewed with disquiet. The main vehicle for the breakout has been the multinational enterprise, bringing its complementary resources of know-how and markets. One type of production has involved components for vertical integration into assembly elsewhere for worldwide distribution. Another has been "horizontal" integration in which the access to markets and trade names has depended on the parent multinational firms.

As some observers see it, this system leaves the poor countries providing low-skilled labor, priced at a competitive world-market rate, while the main rewards for capital, management, and know-how are appropriated by the owners of these scarce but internationally mobile factors. In other terms, labor in highly elastic supply is combined with technology, know-how, and capital in inelastic supply, and the monopoly or oligopoly rents are appropriated by the factors in inelastic supply.

How one regards this model depends on the alternatives. Some advocates of international trade unionism would seek to effect a redistribution of the gain by upgrading labor standards in the developing countries, but they must face the constraint of pricing the labor of developing countries out of the market. Others see possibilities of "unpackaging" the inputs of the multinational firm and gaining for the host country a larger share of the monopoly rents. The possibility can be foreseen that host countries negotiating with multinational firms for the establishment of such factories will try to increase their gains in one way or another by imposing various conditions or even through experimenting with taxation and wage policies.

A reverse possibility also exists, namely, that a host country may seek to obtain a commitment to export a certain part of the output as a condition for the establishment of an import-substituting enterprise. There seem to have been a few such cases in Kenya recently.

If the East African Common Market, particularly in an enlarged form extending beyond the present membership, were ever to come into being, this could of course be the occasion for negotiations of regional industries--but that is another story.

THE CASHEW NUT STORY

One is tempted to say that the experience of Tanzania in developing, choosing, and installing an appropriate technology for the processing of cashews affords, in a nutshell, an object lesson in many of the issues confronting governments and enterprises. The processing of cashew nuts is not exactly the prototype of the operations of a multinational enterprise. Still, it raises many similar problems for host country and foreign participant. The following remarks illustrate just a few of the highlights; they do not do justice to the complexity and drama of this project, which is to be enacted on the very scene of the ill-fated groundnut scheme.

We start with the fact that Tanzania, which produces more than 25 percent of the world's cashew nuts, processes only 10 percent of its crop. The rest is shipped for processing to India and then to markets in North America, Europe, and the Soviet Union, with which India has a special trade agreement. From each ton of raw nuts exported to India, the additional foreign exchange that would be gained by Tanzania if the crop were processed locally would be about Shs. 725 per ton, roughly the margin between the prices of raw nut free on board (f.o.b.), Tanzania raw nut prices and f.o.b. Indian kernel-plus-by-product prices. This represents 50 percent of the present raw nut prices f.o.b. Tanzania. Substantial sums are involved. Even if only one-third of Tanzanian output were processed locally, the additional foreign exchange earnings would be $4.4 million per year in U.S. dollars. *

The first question is, therefore, why does it pay to ship raw cashew nuts from Tanzania to India? The question is all the more intriguing when account is taken that the cost of shipping from Tanzania to India, Shs. 200 per ton, would be eliminated from the cost,

---

*A separate project is the potential for improvement in the yield of cashew trees through an agricultural extension project that is regarded as quite feasible and would be capable of increasing the value of nuts from existing trees by an amount roughly equal to the potential increase from processing. This is, of course, a long-term project, and involves different considerations from the question of local processing.

insurance, and freight (c.i.f.) European price by direct shipment from Tanzania.

Basically the answer is that with both countries using the most labor-intensive technology, hand-processing, the cost of production in India is so low that it pays to ship nuts to India not only from Tanzania but from Mozambique, the world's largest producer. The wages of Indian cashew workers, mostly female, are about one-third the Tanzanian urban minimum wage. Furthermore, Indian workers are more productive because they are motivated by piece rates and lack of alternative employment, including employment in subsistence agriculture.

In short, with unskilled labor in Tanzania paid at the minimum wage of Shs. 145 per hour, it has not been possible to overcome problems of labor supply and labor productivity, using essentially identical technology. Labor costs in a Tanzanian hand-processing unit are estimated at Shs. 504 per ton, at a monthly wage of Shs. 140, compared with Shs. 265 per ton in India.

Since the early 1950s at least, the leading reaction to this state of affairs in Tanzania, and London, in both official and private circles has been to seek a solution through mechanizing the quite complex process of cracking, shelling, peeling, and grading cashew nut kernels. A leading actor has been the official Tropical Products Institute in London, but foreign private firms, both food processors and machinery manufacturers, have been engaged on a large scale. Numerous devices were designed and tested but found deficient.

In the late 1960s a new phase opened. Two highly mechanized, capital-intensive devices were developed and installed by private firms, both in partnership with the official Tanzania Development Corporation. The corporation also was operating a hand-processing project at a loss.

Thus far, both mechanized processes have proved noncompetitive with Indian hand-processing. One, of Italian design, Oltremare, after some years of continuous losses has been able to earn a slight profit only because it receives raw nuts from the government at a subsidized price, that is, below the opportunity cost of shipping the nuts to India. The other, of Japanese design, was apparently constructed with a view to obtaining a high yield of cashew nut shell liquid, a corrosive material that has industrial uses in paints and varnish productions, but at the cost of a lower yield of kernels. The Japanese process has been operated at a loss for several years. Its use was gradually suspended in the early 1970s for redesign.

An important factor explaining Indian competitiveness against the mechanized processes is that the hand-peeling by Indian workers yields a more valuable product. The yield of whole nuts in India is 80 percent of the total compared with 65 percent in the Oltremare

process. Because of the high quality, the Indian processing yields a
higher revenue per ton, which roughly cancels out the cost of shipping from Tanzania to India of Shs. 200 per ton. In order to be competitive, Tanzanian mechanized processing must not exceed 600
shillings per ton--the Indian level--so that the extra revenue earned
by India for higher quality may be offset by the cost of shipping to
India. Similarly, Tanzanian hand processors produce a quality that
allows them to have no more than Shs. 100 higher processing costs
to compete with India. But Tanzania's (Oltremare process) current
costs, excluding capital charges, alone are above India's total costs.
The subsidy in the form of low nut prices has been Shs. 200 below
the f.o.b. export price to India.

In the face of this failure, efforts in Tanzania have shifted to
another form of mechanization, which has been developed by the
Tropical Products Institute. This machine operates with a smaller
intake of nuts and does not carry out as extensive a peeling operation
as the other two mechanized devices. It has been licensed to a
British machinery producer and installed in quantity in Mozambique,
where it is operating at a profit.

With this example, the Tanzanian government is now seriously
considering a major project for mechanized processing of cashew
nuts in the southern province. According to experience in Mozambique and at present wage rates in Tanzania, this process should
compete with the hand-labor process in India. This will be increasingly true as wage rates increase in India and Tanzania over the
years. While it is highly mechanized compared with the hand process,
it is relatively labor-intensive compared with the average capital-
labor ratio for Tanzanian industry, being less than one-half of this
average. The machine uses less capital, and less expatriate staff.
It has lower processing costs and operates in smaller units of
capacity per ton of product processed.

It should be noted that even with a shadow wage rate, that is,
a subsidy, substantially below the Tanzanian minimum wage, which
would have to be paid either by the cashew nut farmers or the government, production by the hand process in Tanzania would be a doubtful
social benefit compared with the newly developed mechanized process
or with Indian hand processing. Judging from the experience with the
new British process, hand processing will become increasingly less
competitive in India as well as in Tanzania. This is true despite the
fact that hand processing provides nearly four times as much employment and machine processing at about one-fourth the cost in fixed
investment and 44 percent of total investment. Even if the problem
of Indian competition did not exist, the mechanized process would be
socially more productive in Tanzania than Tanzanian hand processing.

What moral can one draw from the cashew nut story? First, the story is not over yet, since the cost of operating the newest mechanized process will not be known until it is in operation. However, it seems clear that it is superior both to the labor-intensive and to the large-scale capital-intensive (Oltremare) method. Neither private enterprise nor publicly supported research and development efforts were able to develop a conclusive technique without a long period of trial and error. The beaches are littered with the remnants of unsuccessful designs of cashew-nut processing machinery. But the stimulus of a large-scale export market has been sufficient to mobilize a considerable amount of private and public engineering skill to deal with a technological problem of a developing country.

HIDDEN COSTS OF TECHNOLOGY

The contractual relations between suppliers and recipients of proprietary technology are usually established in circumstances that permit various hidden costs to be attached to the transaction by the supplier in addition to the cost expressed in monetary terms, for example, royalties.

The nature of these hidden costs may depend on the type of contractual relationship. Thus, the awarding of a license to use patented technology may be conditional upon various undertakings restricting exports by the licensee, committing the licensee to exclusive purchase of components from the licensor and so on. At this stage of industrial development in Kenya and Tanzania it is doubtful whether such restrictive business practices impose a major cost on the transfer of technology. As the economy develops, and particularly as export possibilities grow, they may become more serious. Short of international action outlawing or regulating such practices, which cannot be regarded as imminent, the strongest defense a host country can erect against such practices is the enactment and enforcement of antitrust measures and the monitoring of contracts by a government agency to reduce or eliminate their incidence.

In the case of an affiliate of a multinational enterprise, part or all of the cost of technology is implicitly contained in the profit of the local affiliate, even when a nominal royalty or other fee may be charged for the use of a specific technological know-how, but the link between parent and affiliate allows for other transactions that can be priced to alter the accounting profit recorded for local tax purposes by the affiliate. Thus arises the phenomenon of "transfer pricing," particularly the possible overinvoicing of imports of intermediate goods supplied by the parent firm and the charging of various fees to the affiliate for centralized functions of the parent firm that may or may not have any direct relation to services rendered or to be rendered to the affiliate. A major motivation for understating profits in one jurisdiction is to minimize the parent firm's global tax liabilities by taking advantage of differences in national tax rates and regulations; but there are other motivations, including evasion of or anticipation of exchange restrictions on transfers of profits, short-run expectations of devaluation of the host country's exchange rate and, according to some analysts, a desire to show a profile of low profitability in some areas for political or commercial reasons.

Several measures have been taken recently in both Kenya and Tanzania that bear on this problem. The most comprehensive was the institution in late 1972 of a system of monitoring the invoice prices of a substantial part of imports and exports. The immediate motive appears to have been the anticipation of capital flight by resident Asians under pressure of Africanization policies, including the Africanization of business enterprises. The policy has been applied generally, however, and thus covers most transactions of multinational enterprises with their Kenyan and Tanzanian affiliates. An obvious problem is the determination of the "arm's length" or reference price by which the degree of overinvoicing is to be judged. The present system involves control at the port of embarkation that is carried out on behalf of the government by a private agency for an agreed fee.

The knowledge that such a check will be made and that shipment may be blocked if overinvoicing is judged to occur should be a significant deterrent. Whether other simpler and possibly cheaper procedures might be developed to achieve the same effect is a matter for consideration.

Although not intended to deal with the problem of transfer pricing, Kenya's new policy of leveling up import duties on intermediate goods and capital goods that formerly entered duty free should act as a deterrent to overinvoicing. The higher the tariff rate, the less the incentive to take profits through raising prices of such shipments rather than as normal accounting profits of the enterprise. In fact, it is theoretically possible to raise tariffs, while lowering the tax rate on corporate profits, so as to eliminate completely any incentive to overinvoice. But this extreme step, which was advocated by a recent ILO mission to Kenya, would require tariff levels that would distort the tariff structure from the point of view of commercial and industrial policy.

Finally, Kenya recently imposed a 20 percent withholding tax on royalties, management fees, and similar charges paid to parent firms by affiliates. Before this step, Kenya's policy practically ignored such charges for tax purposes, in contrast to many developing

countries, which in some cases partly or wholly disallowed them on
the ground that they should be treated as profits and taxed as such.
It is understood that the rationale of the 20 percent rate is the assumption that, allowing some variable cost in supplying know-how,
this rate corresponds to the equivalent of the 40 percent rate of tax
on corporate profits plus the withholding tax on dividends of 12 percent.

Through these three steps--monitoring prices, duty revision,
and withholding taxes--Kenya has moved significantly to overcome the
problem of transfer pricing. Tanzania has moved in a similar way,
except for tariff policy. The operation of her State Trading Corporation also may contribute. There are more complete solutions, but
they involve essentially some form of international cooperation between governments of the home country of the multinational and the
host countries where their affiliates operate. These include agreement on formulas for transfer pricing and taxation by home countries
on a global basis of consolidated income of multinational enterprises
as earned.

JOINT VENTURES AND MANAGEMENT CONTRACTS

Their official policies toward economic development and toward
foreign investment differ, but Tanzania and Kenya have experimented
increasingly with joint ventures and management contracts as alternatives to the traditional form of a wholly or majority-owned subsidiary
of a foreign enterprise. The basic objective of these policies is to
reduce the element of foreign control and its various costs to the host
country while retaining the foreign technology and management for
which no local alternative is available.

The outcome of such policies will depend on supply and demand
conditions. In negotiations for minority joint ventures or management contracts, and often a combination of the two in a single package, technology is the trump card in the hands of the multinational
enterprise. It is possible to think of a typology of industries with
more or less systematic attitudes regarding concluding agreements
for joint ventures, licensing, and management contracts. Industries
employing "high" technology with a strong possibility of extracting
rents for a long time are likely to resist such arrangements. But a
technological lead is a wasting asset whose value declines over time;
since the cost of using it, once it has been developed, is small, there
is an incentive to sell it before competitors do. The same does not
apply to products that are differentiated not so much by sophisticated
technology as by brand names with supporting advertising expenditure.
Such firms remain in a strong position to hold out against proposals
for minority joint ventures and management contracts from developing

countries. Still another tactic of multinational enterprises as they lose control of technology is to offer a share of the world market to their local subsidiaries through participation in a scheme of international vertical integration.

In short, a country that has embarked upon a policy of replacing controlled affiliates with minority joint ventures, licensing, and management contracts has an almost bewildering array of alternatives to weigh.

Following the Arusha Declaration of 1967, Tanzania's policy has been defined rather rigidly. The nationalization of a wide range of foreign enterprises was only partial, however, being limited to reducing foreign participation to less than 50 percent of the equity. The general policy that was developed after 1967 with respect to future foreign participation in Tanzanian enterprises left room for joint ventures and even, in the category of certain consumers' goods industries, for foreign majority participation.

The primary meaning of the nationalization action in Tanzania and the policy regarding future foreign participation was to force the government, acting mainly through the parastatal National Development Corporation, to develop a policy and a capability with regard to the negotiation of management contracts, most of which in the initial stage were negotiated with the foreign firms formerly owning the enterprise. The pure licensing of know-how by a wholly local enterprise has not developed, because there is little possibility for establishing a wholly indigenous firm that can effectively utilize proprietary know-how on this basis.

Whether used separately or in combination, the management contract and the joint venture have characteristics that should be analyzed carefully by the host country as an integral part of negotiations with the foreign enterprise involved. A sketch of some of the issues is all that is possible here.

The issue of arranging the most suitable form of management contract arises in both Tanzania and Kenya, but more so in the former since the share of foreign equity in Tanzanian enterprises, in the light of the Arusha policy, can be expected to be generally smaller or even zero. The host country's objective is simple: to obtain the most efficient inflow of technological and managerial know-how at the least cost. The bargaining position of the foreign participant was indicated earlier. There has been much discussion of defining the responsibilities and authority of the management agents. Basically, there is little point in trying to outline these matters in detail. The only effective control resides in the competence of the Tanzanian board members and the Tanzanian staff of the enterprise. Control over specific financial transactions and procurement operations is a different matter, requiring more specification in advance.

Formulas to determine the fees paid to management agents
must provide an adequate incentive for efficient performance. Gear-
ing fees to profits seems an obvious answer, but does not take into
account the "opportunity cost" of obtaining management in some circumstances. Much will depend on how much equity, if any, the management agent has in enterprise. If the share of equity is negligible
or small, a fee as a percentage of profits subject to a minimum
would be logical, provided that the percentage of profits should decline as profits rise to some agreed level of output. In some cases,
a lump-sum guaranteed payment may be necessary. Fees based on
sales obviously are least likely to provide the incentive desired, and
should be avoided.

Similar considerations arise with regard to payments of royalties and fees for licensed technology and for the use of brand names.

The duration of a management contract is another dimension to
be considered. There is frequent question whether the training of
local personnel is proceeding fast enough to take over certain functions and also whether it would be cheaper to hire foreign technicians
on an individual contract basis rather than as a team. Clearly a team
represents a more effective capability, particularly if there are continuing links with the foreign enterprise in research and development,
purchasing, marketing, financing, and other functions.

Experience in Tanzania suggests that management contracts
are workable but, like other institutional arrangements, they have
their shortcomings and administrative difficulties that require careful monitoring. Furthermore, experience thus far has been largely
with firms that retained a large equity from a partially nationalized
investment. A few management contracts for new ventures have been
negotiated, but, to a large extent, for standardized products requiring limited technological sophistication and destined mainly for the
local or east African market. Thus, it is premature to judge how
extensively this device can be used effectively as Tanzanian industry
develops further.

The cost/benefit balance of joint ventures can be usefully discussed in the context of Kenya. Kenyan authorities have fostered a
policy of joint ventures with the parastatal Industrial and Commercial Development Corporation (ICDC). The stated objective has been
to achieve "control" and by implication greater benefits to the national economy from such control. The Kenyan involvement has
ranged from majority equity participation by the ICDC to quite small
Kenyan minority participation. Unlike in Tanzania, the policy has
been flexible, and with a few exceptions no forced nationalization,
particularly in the manufacturing sector.

A systematic analysis of the cost/benefit balance of joint ventures in Kenya would have to allow for at least the following considerations:

A management contract or the equivalent may often be the price of agreement by a foreign enterprise to participating in a joint venture as a minority partner. The cost of this arrangement to the host country must be weighed. The degree of effective control that can be exercised in a joint venture by participation in the Board of Directors may vary and, in some cases, be illusory. The question of control raises the issue: control over what for what purpose? It is difficult to appraise it without considering the effectiveness of general controls on the economy like taxation, exchange control, labor legislation, antitrust legislation, and monitoring of transfer pricing. Gaining experience in management through participation on the board deserves consideration; but this is different from Africanization at the managerial level, which can occur with or without local equity participation.

The opportunity cost of the capital invested by Kenya has to be weighed against the possible uses for other projects. It may well be that in some cases, the Kenyan participation was requested by the foreign partner, possibly as an assurance against political uncertainty and also with a view to obtaining preferential treatment in governmental regulations. The business motivations and tactics of the foreign partner must be taken into account. In joint ventures there may be a tendency to take profits through charging higher royalties and prices on intrafirm transactions, whereas in a wholly owned subsidiary royalties often are not even an approximate measure of the "arm's length" cost of technology.

One of the most important differences, potentially, between a wholly owned subsidiary and a joint venture is in the different access to export markets. There is evidence that the joint venture is less likely to be offered access to the channels of distribution of the foreign firm.

In short, the question is whether the joint venture is an effective instrument to accomplish certain objectives of the host country, namely, control of the enterprise in the national interest, a larger share of the gain from the enterprise, or a greater use and development of local resources, particularly human resources in the form of managerial personnel. It would seem that the joint venture should not necessarily be regarded as the major instrument through which these policy objectives are sought.

TOWARD A NATIONAL POLICY ON FOREIGN
BUSINESS PARTICIPATION

It has been sufficiently emphasized in the preceding pages that the elements of a national policy toward direct foreign investment also contain, explicitly or implicitly, most of the elements of a policy toward the importation of proprietary technology. One may use the rather awkward phrase "foreign business participation" to include these two sides of the coin. With this in mind, we can pull together some of the threads we have discussed.

In Kenya the policy toward foreign business participation is increasingly seen, as it should be, as part of the strategy of industrial development. The significant reduction of protectionism that seems to be under way should have a profound impact on foreign and domestic investment. It should encourage greater uniformity in the extension of concessions to foreign investors and, if supported by more systematic and consistent project evaluation than heretofore, lead to a more efficient and less controversial set of industrial projects.

Kenya appears likely to continue to follow a flexible attitude toward joint ventures, management contracts, and similar alternatives to wholly owned subsidiaries. But there is a tendency to scrutinize more carefully all such contractual arrangements and to establish appropriate administrative machinery within the government for this purpose. The basic logic of this is to ensure that negotiations with potential foreign business participants are carried out from a position of maximum strength and reflect national interests in a consistent way. Implicitly or explicitly this should mean the greater use of systematic project appraisal methods, the cost/benefit approach, rather than reliance on pragmatic tests of particular projects such as their direct contribution to foreign exchange or employment. Also, it will mean a less passive attitude toward foreign investors and technology suppliers. With specific projects developed by the local authorities, the negotiations with foreign investors should be more balanced.

To support such a negotiating policy, there will be need of an increased supply of industrial and technical information, permitting a better evaluation of alternative sources of technology and of alternative technologies. Some steps have recently been taken in this direction, but more will be needed.

The recently introduced monitoring of transfer prices also may be considered--together with the introduction of a tax on royalties and management fees, indicating the application of a stricter policy.

In Tanzania, foreign business participation proceeds within a narrower scope and more regulated framework than in Kenya and with different social priorities and attitude toward the private sector. The rigid limits on foreign equity participation are one symptom of this approach. Nevertheless, there is a degree of flexibility in the administration of policies toward foreign business participation. There is the same interest in project appraisal using somewhat different methods from Kenya, and in negotiating with foreign investors and technology suppliers from a position of strength. There is also the awareness that a major requirement for negotiation is a more effective system of acquiring industrial information.

6

THE TRANSFER OF AGRICULTURAL AND FORESTRY "TECHNOLOGY" BETWEEN REGIONAL AND NATIONAL RESEARCH INSTITUTIONS
Theodore W. Schlie

"Technology Transfer" has recently become a widely used phrase that often leads to as much confusion as clarity about what is being discussed, so this chapter will begin by specifying what it is that will be reported here. As differentiated from technology transfer to developing countries, the subject of this chapter is technology transfer within a developing region. The words "region" or "regional" as used here will always refer to a multinational rather than a subnational unit of area or government.

As differentiated from horizontal technology transfer, which often becomes confused with or indistinguishable from "diffusion of innovation," in which the state-of-the-art stays approximately the same but moves across space, national borders, or cultures, the subject here is vertical technology transfer, in which the state-of-the art does change and the technology may or may not move across space, national borders, or attached cultures.

Vertical transfer of technology thus involves the transitions necessary to progress from fundamental research to applied research to development or engineering or both and to application. Although it is generally recognized that African countries will have to import a major share of their technologies from developed countries (horizontal transfer), for the foreseeable future, it also is generally agreed that they should progressively produce and apply more of their own technologies (vertical transfer).

The essential elements for technology transfer are a sender(s) and a receiver(s). "Technology," as used in this chapter, may refer to basic scientific results and/or information as well as to machinery or other forms of hard technology. The technology, the linkage mechanism between receiver and sender, and the socioeconomic and cultural environment surrounding the transfer are the other three elements often cited in the literature on technology transfer. A simple model of these elements is shown in Figure 6.1.

FIGURE 6.1

Technology Transfer Model

Socio-Economic/Cultural Environment

Linking Mechanism

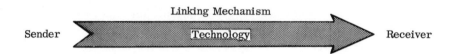

Sender   Receiver

The discussion and results in this chapter are based on the author's doctoral dissertation research carried out in east Africa in 1971-72 and completed in August 1973.[1] This study focused on the interplay between the East African Agricultural and Forestry Research Organization (EAAFRO) of the East African Community (EAC) and the national agricultural and forestry research systems of the partner states--Kenya, Uganda, and Tanzania. The basis for almost all theories of regional or multinational cooperation involves some aspects of the economies of scale and increased capabilities that can be derived from a centralization of effort. This occurs in particular when nations are small and fragmented and relatively lacking in scarce resources. Recently the arguments for regional economic or political cooperation or both to overcome such obstacles to development as small national markets have been increasing, both within developing countries and from national and international aid agencies. This is particularly true for Africa, which has a large number of relatively small, poor countries.

> The unsatisfactory outcome of the first development decade (which has occasioned almost worldwide disappointment) is now giving rise to an awareness that most of the under-developed countries are too small to be able to create modern, viable industrial structures within their national frameworks. . . . Hence the necessity for such countries to follow the example of the European Economic Community in pooling their resources

and coordinating their development in a movement
towards regional solidarity.[2]

Someone has remarked that Africa has the highest
rate of frontiers to total area of any continent.
Whether this is so or not, it certainly contains a
number of states which are very small in terms of
population and natural resources . . . decoloniza-
tion has so far resulted in the fragmentation of an
already over-divided continent.[3]

To the extent that progress depends on the division
of labor and consequently on the size of market,
reasonable market size is a concomitant of reasonably rapid development. This is especially true in
the light of African economic nationalism, which,
by and large, obstructs economic integration into
larger units. Individual African markets are now
extremely small. . . . The general smallness of
African markets and their natural complementarity
are necessary and sufficient economic reasons for
urging that African states (with possible exceptions
such as the U.A.R. or Nigeria) combine their efforts, wherever possible, to attain greater economic
viability.[4]

Analogous arguments also have appeared recently for the creation
of regional research institutions in developing countries. In its
third report to the Economic and Social Council of the United Nations, the Advisory Committee on the Application of Science and
Technology to Development (ACAST) stated the following:

An important goal for developing countries is the
achievement of a substantial degree of scientific
and technical independence, based on adequate and
vigorous national institutions. Until that stage can
be reached in some countries, regional institutes
must be relied upon. . . . Many of the problems
of developing countries to which science and technology can contribute are regional. Moreover, the
minimum effective size of the needed research activities, both as to staff and special equipment,
may make it necessary for countries in the region
to pool their resources and personnel for combined
action. Such pooling may take the form of a regional

> institute, or it may consist of a consortium of
> national institutions closely linked for coopera-
> tive action.[5]

Unfortunately, most of the past attempts to implement general
regional economic or political schemes or both, or specific regional
scientific research schemes in developing countries, have not succeeded in reaching their objectives. The Central African Common
Market, Maghreb Common Market, Latin American Free Trade
Association, Central American Common Market, Regional Cooperation for Development, West African Customs Union, among others,
are some regional schemes that either have ceased to exist or failed
to meet expectations. There have been fewer indigenous attempts to
establish and maintain regional scientific research institutions, but
here again the record has not been encouraging. An entire set of
West African Cocoa, Palm Oil, and other research institutes established in the British colonial period, for example, became national
institutes of Ghana, Nigeria, or other west African nations shortly
after independence. Therefore, the viability of regional scientific
research institutions in developing countries is questionable.

In east Africa, however, there is a viable general economic
and political regional organization--the East African Community--
that has, despite repeated threats to its existence, continued to exist
and develop. One of the EAC's activity areas is scientific research.[6]
The set of 12 regional EAC research institutions includes EAAFRO
at Muguga, Kenya; the East African Veterinary Research Organization (EAVRO), also at Muguga, Kenya; the East African Marine
Fisheries Research Organization on the island of Zanzibar; the East
African Freshwater Fisheries Research Organization at Jinja, Uganda;
the East African Institute for Medical Research at Mwanza, Tanzania;
the East African Virus Research Institute at Entebbe, Uganda; the
East African Trypanosomiasis Research Organization at Tororo,
Uganda; the East African Malaria and Vector Borne Disease Research
Institute at Amani, Tanzania; the East African Leprosy Research
Center at Alupe, Uganda; the East African Tuberculosis Investigation
Center at Nairobi, Kenya; the East African Industrial Research Or-
ganization at Nairobi, Kenya; and the Tropical Pesticides Research
Institute at Arusha, Tanzania.

EAAFRO, however, is the oldest and largest of these institutions, reflecting the importance of agriculture in the region. All
three partner states also have established counterpart agricultural
and forestry research systems with which EAAFRO is to cooperate.
Thus, there were defined organizations at the regional and national
levels of agriculture and forestry research in east Africa, and the
study focused on the relationships between them.

The general research questions that this author addressed
were: "How is the regional research institution of EAAFRO relating
to national agricultural and forestry research systems in Kenya,
Uganda, and Tanzania?" and "What are some of the factors causing
and/or influencing those relationships?" As part of an overall regional framework that once was to have become an East African
Federation, EAAFRO can be perceived as an integrating agent--suc-
cessful or otherwise--for agricultural and forestry research in east
Africa. It therefore was felt that theories of integration might offer
some insights into the relationships between these regional and national research systems. The study, therefore, attempted to take
some aspects of general political and/or economic integration theory
and to adapt and apply them to the specific and specialized activity of
agricultural and forestry research in the specific regional setting of
east Africa.

Three major exercises were encompassed in the study: a
methodology for "objectively" measuring the scientific benefits from
EAAFRO and determining how those benefits were distributed among
the partner states was developed and used;* transaction analyses of
the professional visits of EAAFRO research officers to locations in
the partner states and of the visits of national level representatives
from the partner states to EAAFRO headquarters were performed;
and perhaps the greatest amount of research effort went into an exploration of the reaction-response opinions of national researchers
toward EAAFRO. The attempt was to comprehend what these re-
spondents immediately felt about EAAFRO, rather than what they
might "rationally" conclude after going through some internalized
reasoning process. A number of propositions were tested in this
context.

Some information for this study was gained from EAAFRO and
national publications, but by far the most important source was extensive interviews with EAAFRO and national research officers. In
late 1971, 37 of 40 EAAFRO research officers were interviewed in
depth about all the different research projects or service activities
on which they were working. This included aspects of collaboration
with the partner states, field sites in the partner states, potential

*In addressing the issue of benefits from a scientific research
institution, it was found to be helpful to distinguish between research
results and scientific services. For the purposes of the study, "research" was characterized by its experimental nature and by work
done for one's self; "services" were also scientific in nature, but
usually involved routine work done for someone else's research.

applicability in the region, and communications patterns with the partner states. At the national level, 165 out of approximately 220 research officers in east Africa* were also interviewed in depth on their reaction-response opinions toward EAAFRO and the experiences and interactions they had had with EAAFRO that were hypothesized to relate to those reactions and opinions or both. The full results of the study are discussed in the author's dissertation. Only those aspects relating to technology transfer will be discussed here. Figure 6.2 shows the spatial distribution of the national research stations visited.

The primary official relationship between EAAFRO and national agricultural and forestry research systems is vertical technology transfer from the regional to the national level. This formal functional relationship was first stated by B. F. Keen, director of EAAFRO at the time it moved from Amani in Tanganyika to Muguga in Kenya just after World War II. In the following quotation, "Research Organization" refers to EAAFRO:

> It is desirable, however, to set out clearly the respective functions of the Research Organization and the territorial Departments in the three phases of the sequence. The first phase, basic research, is normally the responsibility of the Research Organization, because the men engaged on it need to be physicists, chemists, plant physiologists and the like rather than agriculturalists. The third stage is, obviously, the responsibility of the territorial Departments, whose administrative organization is designed, and whose agricultural, forestry and veterinary officers are trained, for the task of introducing proved and tested results into practice. To do this, they must have available the results from the second or technological stage. Similarly,

---

*This total includes 75 out of 107 (70 percent) from Kenya; 44 out of 50 from Uganda (88 percent), and 46 out of 63 (71 percent) from Tanzania. With one exception, the national researchers not interviewed were missed for random reasons. For example, they were on a field trip when the interviewing took place, and so their absence should not have interjected any systematic bias into the results. The one exception was forestry researchers from Tanzania, who could not be interviewed because permission to conduct the research was not granted by the Tanzanian Forest Department.

FIGURE 6.2

Locations of National Agricultural and Forestry Research
Institutions Visited in East Africa

the Research Organization must carry any results
of its basic researches into the technological stage,
otherwise these results might remain little more
than scientific curiosities. Therefore, as the Departments and the Organization are equally interested in the technological state, there are many
opportunities for joint investigation and teamwork
by the specialist officers in the Departments and
the research staff of the Organization. [7]

The same theme was echoed in 1952 by E. B. Worthington, scientific
secretary to the East African High Commission:

It is, however, possible to make a broad definition
of the functional relationship between the regional
organization and the territorial department working in the same general subject. There are three
main stages in science; firstly, there is fundamental or long-range research which is designed to
discover new principles; secondly, there is the
technological stage in which a new principle is
tested in a variety of local conditions in order to
determine how far it is applicable and to adapt the
technique to local circumstances; thirdly comes
the application of the new principle in farming,
medicine, industry or whatever the subject may
be. The regional research organization is concerned with the first two, the fundamental and
technological stages; it is not usually concerned
with the application to practice. The territorial
departments, on the other hand, are primarily
concerned with the application of the new knowledge
and all the executive work which is thereby entailed, but it has also to take a large share in the
technological stage of the trial in local conditions.
It is thus in the second or technological stage that
the functions overlap and there is need for the
closest collaboration.

This distinction of function has, of course, many variants. [8]
By 1961, the report of the Frazer Commission, the recommendations
of which served as the basis for the present regional research administrative setup, perceived the functional relationship between regional and national research as follows:

> East African research is organized on the basis of
> interterritorial and territorial research groups.
> They are complementary to each other and it is
> essential that they should work together. Both
> groups may undertake short-term and long-term
> work . . . in general, the territorial research
> groups are more concerned with immediate prob-
> lems and give invaluable assistance in field work
> to the interterritorial research groups, which
> should concentrate their efforts mainly on longer-
> term studies. The territorial research services
> also provide the link with the extension services
> which are responsible for the application of the
> research findings.[9]

A 1966 publication on Research Services in East Africa stated:

> The principal functions of EAAFRO are to under-
> take research work that is either longer-termed
> than is easily undertaken by the Territorial De-
> partments of Agriculture and Forestry, or which
> requires highly specialized equipment or Research
> Officers, which can only be justified in a central
> East African laboratory. EAAFRO is only con-
> cerned with problems of local interest when spe-
> cially invited by a Territorial Department and its
> function then is to supplement and supply the back-
> ground to the work of the territorial research
> workers.[10]

A recent annual report of EAAFRO includes most of the above functional role description, but adds the following:

> . . . EAAFRO is not an advisory organization as
> such, though scientific advice and guidance are
> readily given by the research staff on request to
> national and other research workers. . . . Re-
> search at EAAFRO is as much concerned with the
> solution of problems as it is with the acquisition
> of new knowledge, and also that much of the re-
> search consists of projects carried out with the
> cooperation of other research institutes.[11]

There has been, therefore, a certain amount of concern shown for
the research relationships between EAAFRO and national agricultural

and forestry research systems. There also has been some ambigu-
ity and confusion about what role EAAFRO should play in relation to
national systems, and what would be the converse. Through the
years, however, the basis for the existence of regional and national
institutions in the same functional area of agricultural and forestry
research has been that the regional institution would do the more
fundamental, and more expensive, more specialized, research and
the national institutions would do the more applied research. Some-
where in between there would be an overlap of duties and a transfer
of results. An implicit and necessary condition for such a relationship is that both levels of research must be working on the same, or
complementary, projects.

As part of the basis for that exercise that sought to explain the
immediate and general reaction-response opinions of national researchers toward EAAFRO, one of the propositions tested was as
follows:

> The degree of positiveness of the general reaction-
> response opinions of national researchers toward
> EAAFRO will be directly related to their percep-
> tions of the benefits they have received from
> EAAFRO in the past--i.e., to the amount of Re-
> search Results and/or Scientific Services and/or
> Experimental Collaboration and/or Institution-
> Building and/or Interpersonal Relations and/or
> Publishing benefits, dependent in some cases on
> an External Scientific Aid parameter as described.

In investigating the extent to which these potential benefits
from EAAFRO had actually been used by national researchers, one
finds a number of the results directly related to technology transfer.
First, national researchers were asked for instances in which they
personally used the results of EAAFRO's experimental research in
their own research, and if so, how useful those results were. The
responses are shown below in Table 6.1. All of the results were
also broken down into citizen and expatriate categories for the national researchers in each partner state, but the patterns presented
did not exhibit any significant differences in this case.

The most striking result shows only 48 out of 165 national researchers had ever used research results from EAAFRO in their own
research work. Half of those 48, however, had found those results
to be of "Very Great Use" to them. Only four found the results to be
of "Little or No Use." In the latter cases, the EAAFRO results had
not worked for the national researcher in his own work. In the few
cases where national researchers had used EAAFRO results but

couldn't venture an opinion as to their usefulness, the national projects were just under way and it was too early to judge the usefulness of the EAAFRO input. Most of the utilization cases in the top three usefulness categories were from EAAFRO's soils research--soil physics/agrometeorology, soil chemistry/fertility, and methodologies for soil analyses. A significant number of utilization cases also appeared for EAAFRO's maize research, with lesser numbers in fields like animal production, forestry, and coffee.

TABLE 6.1

National Research Utilization of EAAFRO Research Results

	Number of National Researchers			
Response	Kenya	Uganda	Tanzania	Total
Very great use	13	10	1	24
Much use	4	3	3	10
Some use	2	2	2	6
Little use	--	1	2	3
No use	--	1	--	1
No answer	2	--	2	4
Subtotal	21	17	10	48
Not used EAAFRO results	54	27	33	114
No answer	--	--	3	3
Total	75	44	46	165

A second benefit directly related to technology transfer was experimental collaboration--cooperative situations in which EAAFRO and national researchers were actively working on joint projects. The results from asking national researchers for past or present instances of this nature are shown in Table 6.2. Again, the most striking result shown is that only 35 out of 165 national researchers had ever collaborated with EAAFRO researchers on joint experimental projects. Almost all collaboration occurred in the fields of maize breeding, forestry, soil chemistry/fertility, plant pathology, and sugar cane breeding.

The third benefit related to technology transfer was called Scientific Services. The utilization of eight specific services that EAAFRO offered was investigated: the East African Literature Service, the joint EAAFRO/EAVRO Library, the Plant Quarantine Service, Statistical advisory services, the Herbarium's plant identification and information service, the identification and information

TABLE 6.2

National Experimental Collaboration with EAAFRO

Responses	Number of National Researchers			
	Kenya	Uganda	Tanzania	Total
Past or present collaboration	16	14	5	35
Supervision of VTCs only*	5	2	2	9
No collaboration	54	28	36	118
No answer	--	--	3	3
Total	75	44	46	165

*In these cases, the national researcher only supervised the trials of new plant varieties developed by EAAFRO located at their stations. These trials were conducted at various locations throughout east Africa to test the new variety under local conditions. If cooperative research activity in carrying out these trials was involved, then this was judged to be collaboration; if supervision of the trial sites was all that was involved, this was judged not to be collaboration but these cases are put in a column by themselves.

services of insect, virus, and fungus Reference Collections, the Chemical Analyses service for soil and plant specimens, and the Machinery Coordination Service. The result from asking national researchers about the extent of their use of these services and their usefulness are shown in Tables 6.3, 6.4, and 6.5. The East African Literature Service was the most heavily utilized EAAFRO service, and it was utilized proportionally in all three partner states. One of the reasons for this widespread use was that this service is transferred through the mail. Other services, like the Library of Statistical Advisory Services, require face-to-face contact for transfer to occur, and proximity is thus a negative factor affecting parts of Tanzania and Uganda. Significant alternative sources of several of these services were also found in Uganda and Kenya. Makerere University, the oldest and most capable institution of higher learning in east Africa, was an important alternative source of several scientific services to national researchers in Uganda, where it is located. The National Agricultural Laboratories of Kenya's Ministry of Agriculture, the single largest and most capable scientific research institution in east Africa, was a significant alternative source of several scientific services to national researchers in Kenya. Some national researchers in all three partner states, however, emphasized that they did need particular services and inquired how they might obtain them from EAAFRO. This occurred particularly for statistical advisory services, for chemical analyses, and for agricultural machinery advisory services.

TABLE 6.3

National Utilization of Specific EAAFRO Services

EAAFRO Service	Number of Researchers Who Have Utilized From			
	Kenya	Uganda	Tanzania	Total
Literature Service	53	34	33	120
Library	39	10	4	53
Plant Quarantine	18	11	11	40
Statistical Advice	23	10	5	38
Herbarium	23	5	6	34
Reference Collections	12	5	9	26
Chemical Analyses	10	4	5	19
Machinery Coordination	--	2	--	2

In the statistical analysis of the results, the proposition stated earlier was firmly rejected--that is, the potential benefits, either singly or in total, that a national researcher had received from EAAFRO, including research results, experimental collaboration, and scientific services--did not appear to significantly relate to his reaction-response opinions toward EAAFRO.* In discussing this result, the following comments were made:

> The overall lack of influence which potential EAAFRO benefits appear to have on general opinions toward EAAFRO may not be so surprising if the results from the previous chapter are reviewed. Despite the official pronouncements of the functional relationships between regional and national research, which implies

---

*This rejection was true for the total sample of national researchers and also for citizen, expatriate, and national subsamples. Chi square frequency distribution tests were used in all cases, where the criterion for accepting or rejecting the relationship being tested was if the chi square was significant at a level of .05 or less.

TABLE 6.4

Perceptions of Usefulness of Specific EAAFRO Services--High Users

EAAFRO Service		Number of High Users Who Thought Service was of					
	Very Great Use	Much Use	Some Use	Little Use	No Use	No Answer	Total
Literature Service	45	6	3	3	--	--	57
Library	12	--	1	--	--	1	14
Plant Quarantine Advice	15	2	--	--	--	--	17
Statistical Advice	4	1	1	--	--	--	6
Herbarium	10	--	--	--	--	--	10
Reference Collections	2	--	--	--	--	--	2
Chemical Analyses	1	1	--	--	--	1	3
Machinery Coordination	--	--	--	--	--	--	--

TABLE 6.5

Perceptions of Usefulness of Specific EAAFRO Services--All Users

EAAFRO Service		Number of Users Who Thought Service was of					
	Very Great Use	Much Use	Some Use	Little Use	No Use	No Answer	Total
Literature Service	76	21	15	5	2	1	120
Library	28	7	8	5	--	5	53
Plant Quarantine	29	4	2	2	--	3	40
Statistical Advice	18	5	6	7	1	1	38
Herbarium	26	3	4	--	--	1	34
Reference Collections	15	1	5	1	1	3	26
Chemical Analyses	7	4	4	1	2	1	19
Machinery Coordination	--	--	1	--	--	1	2

more fundamental and more applied research at the
respective different levels in the same field and the
transfer of results in between, the results clearly
show that this in fact occurred only in a limited
number of cases, maize breeding being the best
example. Much of the research being done at regional and national levels--fundamental or applied--
is in different fields, and so there is little national
utilization of EAAFRO research results and little
experimental collaboration between them. This is
perhaps best shown in the results of the Scientific
Complementarity with EAAFRO variable in which
relatively few national researchers fell into the
High Category for the extent to which EAAFRO's
scientific work complemented their own. . . .*

Therefore, it appears that a situation exists
in East Africa in which the benefits from EAAFRO
do not generally flow to national agricultural and
forestry research systems. [This possibility was
recognized earlier in the study. See Figure 6.3.]

There are exceptions of course, but instead of
working together on agricultural and forestry problems, the regional and national levels of research
are working separately. This situation obviously
runs counter to the official pronouncements of how

---

*"Information gained from the national researchers' descrip-
tions of their own work was used to rank how much EAAFRO's work
complemented their own. Work on the same agricultural product--
e.g., plant breeding--were utilized as indicators of complementarity,
and on this basis all of the national researchers interviewed were
categorized in terms of High, Medium, or Low Scientific Complementarity with EAAFRO. The results are shown below in Table 6.6.
The relatively high number of researchers in all three Partner
States in the Low Category can perhaps be explained by the fact that
large numbers of national researchers are working on cotton, wheat,
sesame, potatoes, horticultural crops, etc.--crops that are basically not dealt with by EAAFRO. In these cases, there seems to be
a complete division of labor along agricultural product lines, rather
than the functional division described earlier where EAAFRO does
the more fundamental research and the Partner States do the more
applied. Only in the case of maize, sugar cane, and some disease
aspects of legumes do active, complementary EAAFRO and national
research programs exist." (Schlie, op. cit., p. 248.)

FIGURE 6.3

Input–Output Flow Diagram of Agricultural and Forestry
Research Systems in East Africa

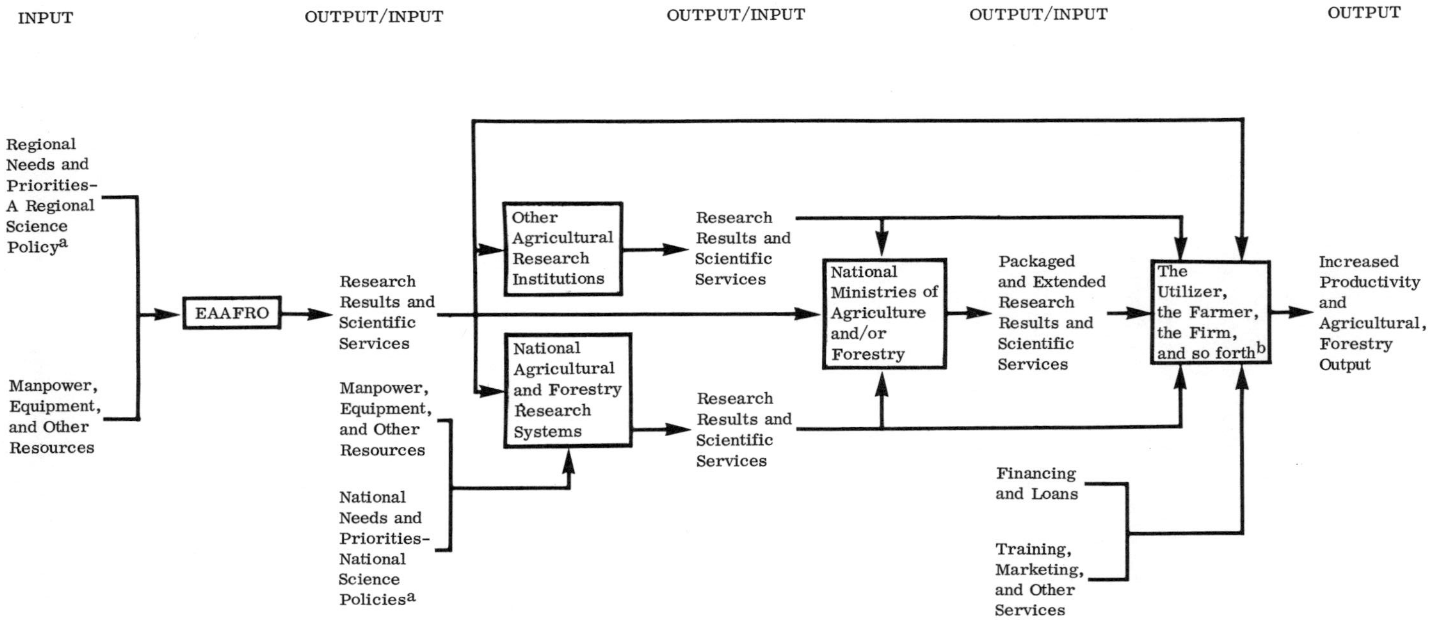

INPUT | OUTPUT/INPUT | OUTPUT/INPUT | OUTPUT/INPUT | OUTPUT/INPUT | OUTPUT

[a]There were no explicit national or regional science policies in east Africa at the time of the study, but the lack of such policies is, in itself, a kind of science policy.

[b]In some cases, the utilizer may be the ministry itself or an institution which is part of or comes under the ministry.

EAAFRO and national agricultural and forestry
research systems are supposed to be working
together, and brings into question the centralization of efforts argument that lies behind EAAFRO
and other regional institutions. Indeed, if a regional institution is not going to handle the entire
job or function that it is set up to accomplish and
national institutions are established to do the
same types of jobs or functions, then some analog
of the "more fundamental research at regional
level and more applied research at national level"
probably exists to justify or rationalize the existence of comparable institutions which might be
accused of duplicating efforts.[12]

TABLE 6.6

The Scientific Complementarity of National Researchers
to the Work of EAAFRO

Responses	Number of National Researchers			
	Kenya	Uganda	Tanzania	Total
Scientific complementarity				
High	24	11	8	43
Medium	18	10	11	39
Low	33	23	27	83
Total	75	44	46	165

In observing that, with the exception of the East African Literature Service, the different scientific services that EAAFRO offers
also are effectively reaching only a limited number of national researchers, I concluded that if national needs are not being met, or
if a service is not being performed satisfactorily by a regional institution, national alternatives are bound to develop. To the extent
that a particular service can be performed more easily or efficiently
at the national level, this is an entirely reasonable prospect. To the
extent that a particular service can more logically be performed at
a centralized and specialized regional institution, however, this may
be an undesirable development. The Literature Service is a good
example of what a regional institution can do to benefit the entire
region when it initiates, develops, and obtains adequate resources

to carry out an active scientific services program. To date, however, an equal amount of concern and effort does not seem to have
gone into the planning and operation of other EAAFRO services.
Other pertinent comments on the transfer of scientific services were
as follows:

> Although some scientific services are more related
> to some agricultural research fields than others, in
> general scientific services can provide aid across the
> board--e.g., the East African Literature Service
> provides scientific information in all fields of agri-
> cultural research. The output from scientific re-
> search, on the other hand, even if it is of a more
> fundamental nature so that the results can be utilized
> in further applied research, is of direct use only to
> other researchers in the same field. Therefore, it
> seems appropriate that EAAFRO might consider the
> emphasis it places on providing scientific services
> to national research systems since 1) with the ex-
> ceptions of the East African Literature Service and
> the Plant Quarantine Service, it does not appear that
> much time or effort has been expended on them;
> 2) at least some of the scientific services already
> offered on a limited scale are urgently needed and
> would be greatly appreciated by national researchers;
> and 3) it appears that the provision of across-the-
> board scientific services might be a very appropriate
> function for a regional institution like EAAFRO to
> carry out. Unlike scientific research, no Partner
> State could claim that these service benefits were
> intended more for one State than another. Problems
> of providing some scientific services over large
> distances would still remain, but much more could
> be done in this area. If one wished to speculate on
> the future and wished to emphasize this point to the
> extreme, one might expect that as national agricul-
> tural and forestry research systems inevitably be-
> come stronger that EAAFRO might find its logical
> role in the region becoming more of a provider of
> expensive but routine scientific services and less of
> a doer of research. Such a development would
> totally change the character of EAAFRO, however,
> since top-quality researchers want to do experi-
> mental research, and it is by no means certain that
> this is desirable. [13]

In attempting to follow the utilization of EAAFRO's research results and scientific services by national researchers, the study utilized the model shown in Figure 6.4. A fatal breakdown could occur in any of these process steps and negate everything correctly done before or after. Perceptions of national researchers about EAAFRO's performance of these three processes, along with alternative explanatory variables of the capacity of the national research institution to absorb and utilize EAAFRO outputs, the extent to which EAAFRO's scientific work complemented that of the national researcher, and the extent of the national researcher's experience in east Africa, formed the basis for the following three propositions that were tested in the study:

FIGURE 6.4

The Important Processes of Any Scientific Research
Institution and Their Intended Results

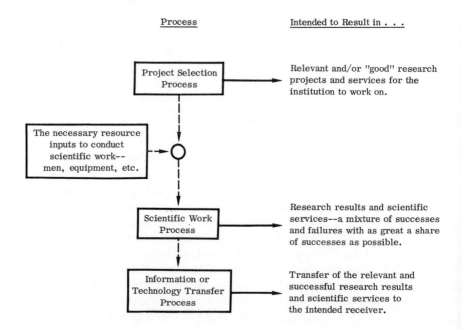

The utilization of the EAAFRO output benefits of
research results and/or scientific services by
national researchers will be directly related to
their perceptions of the relevance of EAAFRO's
projects and services and/or the competence of
EAAFRO's scientific work and/or the success of
EAAFRO's information/technology transfer effort.

The utilization of the EAAFRO output benefits
of research results and/or scientific services by
national researchers will be directly related to
the extent of their experience at their national
station and/or to the extent that EAAFRO's scientific work complements their own.

The utilization of the EAAFRO output benefits
of research results and/or scientific services by
national researchers will be directly related to the
capacity of the national institution of which they
are a part to utilize those results and/or services.

The only aspects of these propositions significantly supported
(p. $\leq$ .01) were, first, that the utilization of EAAFRO research results was significantly related to the extent EAAFRO's work complemented that of the national researcher, and, second, that the utilization of EAAFRO scientific services was significantly related to the
extent of the national researcher's experience in east Africa. Although they were not explicitly hypothesized, a number of additional
relationships were tested for statistical significance. It turned out
that although the national researcher's perceptions of the success of
EAAFRO's information or technology transfer effort to himself
and/or his colleagues at the national level were not related to his
actual utilization of EAAFRO's results or services, those perceptions were directly and significantly (p = .0080). Although an individual national researcher might not have been able to recount any
specific cases of his own utilization of EAAFRO's results and/or
services, he seemed to believe that they were successfully being
transferred to other national researchers, and this belief was related
to a favorable opinion toward EAAFRO. Neither their perceptions
about the relevance of EAAFRO's projects or services or the competence with which EAAFRO carried out its work were significantly
related to those general reaction-response opinions.

Table 6.7 shows the responses of national researchers to the
question of how successful--on a five-point scale--they perceived
EAAFRO's technology or information transfer to themselves, that
is, to the national research systems, to be. Most responses fell
into the three middle categories. It would be difficult to conclude

from these results that there was significant dissatisfaction on the
part of national researchers with the transfer of EAAFRO results to
themselves. It is apparent, however, in examining their response
patterns for perceived relevance and perceived competence, that
perceived transfer success is significantly lower. This relatively
lower pattern was also consistently maintained by all partner states
and by both citizen and expatriate researchers. In arriving at these
responses, moreover, many national researchers internally compared what they perceived EAAFRO's success to be in fulfilling this
function with the success of their own national systems--a comparison that probably put EAAFRO's transfer performance in a somewhat better perspective.

TABLE 6.7

Perceived Success of Results Transfer

| | Number of National Researchers | | | |
Responses	Kenya	Uganda	Tanzania	Total
EAAFRO results transfer perceived to be				
Very successful	4	2	2	8
Successful	22	16	15	53
Just above average	28	17	14	59
Unsuccessful	14	9	8	31
Very unsuccessful	4	--	1	5
No answer	3	--	6	9
Total	75	44	46	165

In explaining their responses to this question, national researchers made many more negative remarks about EAAFRO than
positive, even when answering favorably. Most thought of their response in terms of publications, enough or not enough, rather than
personal contacts. Twenty-one national researchers thought, in
general, that EAAFRO adequately published its results. An additional 29 national researchers favorably referred to the East African
Agricultural and Forestry Journal in this context, an additional 15 to
EAAFRO's Annual Reports, and an additional 11 to the EAAFRO
Newsletter. These responses were all reasonably well distributed
among the partner states.
    Many national researchers felt differently, however. Fifty-five
national researchers--38 from Kenya, 10 from Uganda, and 7 from

Tanzania--indicated in responding that the publication of EAAFRO results was inadequate or that they did not see the results of EAAFRO's work. Why so many national researchers from Kenya compared to Uganda or Tanzania should fall into this category is hard to understand. It might be that being located closer they expected more from EAAFRO than researchers in the other partner states. A few criticized specific EAAFRO publications in explaining their responses. Eight national researchers cited the Journal for being two to three years out of date in publishing results, for publishing low-quality papers, or for publishing only in limited fields. A few cited the Newsletter for not being technical or detailed enough, and a few cited the Annual Reports for being out of date and too long and detailed. All three publications were criticized by a few national researchers for arriving infrequently or not at all, but it was admitted by some that this might be a national problem rather than EAAFRO's. In fact, in recognition of the fact that a "transfer" implies a receiver as well as a sender, two researchers from Kenya, one from Uganda, and two from Tanzania blamed themselves and their colleagues for their unfavorable opinions about the transfer of EAAFRO results.

Closely related to the above unfavorable comments from EAAFRO's transfer efforts was another explanation given by 16 national researchers. Their point was that although there were some EAAFRO publications transferring results, EAAFRO essentially played a passive role in the transfer process. If a national researcher made an inquiry or asked about specific results, he would most likely get a reply, but EAAFRO was not pushing its own results. Although the important factor of personal contacts in the transfer process was not mentioned as much as publications, 13 national researchers did refer to the lack of such contacts in explaining their response. Only two cited the existence of such contacts in their reasons for a favorable response. Specific fields of EAAFRO research were referred to in a few cases as examples where successful transfer had taken place. Those most frequently mentioned were maize and armyworm research. One characteristic these fields had in common was their almost total support by foreign aid sources and the provision in that support for conferences, seminars, and other transfer mechanisms.

This chapter will conclude with discussions of some specific barriers to technology transfer that were evident in the east African context, and some specific mechanisms that were utilized by EAAFRO for technology transfer.

One important barrier in east Africa was the "locale specific" nature inherent in much agricultural and forestry research. This barrier is principally composed of ecological factors, where research carried out on a particular crop at a particular altitude with

a particular rainfall pattern and under particular soil conditions
might not be applicable a few miles away where the altitude, rainfall
pattern, and soil conditions are different. East Africa is a region
that comprises a vast range of ecological characteristics, and much
of EAAFRO's research suffered from this limitation in terms of its
regional applicability.

The best-known case of this among both regional and national
researchers was maize breeding. EAAFRO's maize genetics division was located at a decentralized location at Kitale National Agricultural Research Station in Kenya, where Kenya's national maize
breeding and testing program was also located. Proximity and
natural interchange between EAAFRO and Kenyan programs led to
excellent technology transfer between them, but transfer to Tanzania
and Uganda was less successful. Not only did distance and communications problems exist, but the new varieties developed at Kitale
were developed in consonance with and for the high altitude conditions
around Kitale and other parts of Kenya's "white highlands."* Similar conditions existed in very few places in Tanzania, and not at all
in Uganda. Since maize was the major indigenous food crop in east
Africa--less so in Uganda than Kenya and Tanzania--this discrepancy
was an important ecological and psychological barrier to the transfer of EAAFRO maize technology to those countries.

Even under different and less than optimal ecological conditions, EAAFRO's new maize varieties would probably perform better
than indigenous varieties utilized in Tanzania and Uganda. Therefore, efforts are continuing to test new varieties under different
conditions throughout east Africa in a system of variety trials cen-
ters, and to induce all partner states to support their use by the
people. A mechanism implemented recently by EAAFRO in Tanzania and Uganda to improve the transfer process is the field trials
officer.† These EAAFRO officers are permanently assigned to the

---

*The "white highlands" generally refer to those areas of Kenya
that, in colonial days, were reserved for European settlement,
rather than to any generalized physical feature. Nevertheless, the
British reserved those areas most suitable for agricultural production and those areas generally were in the higher altitudes where
land was more fertile and rainfall more plentiful. This then brings
up the issue of EAAFRO's colonial history and ties to European
agriculture in Kenya, and the effects this is still having on technology
transfer and other EAAFRO-national relationships.

†This mechanism is actually heavily supported by USAID--as
is EAAFRO's entire maize genetics program--and the officers are
U.S. citizens. Their duties include other EAAFRO cereal-breeding

national Ministries of Agriculture in Tanzania and Uganda and live there. Their specific objective is to transfer the maize technology developed by EAAFRO into use through the testing of new varieties under local conditions and the promotion of their utilization by the local people.

Although it is too early to judge the long-term impact of these EAAFRO officers, national researchers in Uganda and Tanzania held very favorable opinions toward them. Table 6.8 shows their responses. Not only was national researcher opinion overwhelmingly favorable toward the field trials officers, there was also extremely high awareness of these relatively new positions. The few unfavorable opinions were related to feelings that this position was only duplicating work already being done, or that an east African citizen should have been recruited to fill the position since advanced scientific qualifications were not required.

TABLE 6.8

General Opinions Toward EAAFRO Field Trials Officers

Responses	Number of National Researchers from		
	Uganda	Tanzania	Total
Field trials officers perceived			
Very favorably	20	20	40
Favorably	11	11	22
Neutral	5	4	9
Unfavorably	3	--	3
Very unfavorably	1	1	2
No answer	4	10	14
Total	44	46	90

Although there are a few Kenyan national research institutions located close to EAAFRO, once one leaves the immediate vicinity around Nairobi, distance and communications become an increasing barrier to technology transfer from EAAFRO. As is the case in many developing countries, transportation and communication linkages are very weak in east Africa, especially once the principal

programs for sorghum and millet, but these latter cereals are of minor importance compared to maize.

roads are left behind. Proximity was found in the study to be significantly related to personal contact transactions between EAAFRO
and national research officers (p = .0014), one of the principal mechanisms by which technology transfer occurs. Although telephone
communication was possible--even if unreliable--within the larger
metropolitan areas of the region, delays and breakdowns over longer
distances made this form of communication virtually unusable. Distance interacted with national borders to produce other barriers to
a form of communication like the postal service. Not only were
there normal bureaucratic delays that resulted from, for example,
the procedural routing of foreign mail through Ministry headquarters
channels, but in recent years tensions and hostilities between partner states have resulted in partial breakdowns of all communications.
In many cases these tensions were related to economic issues when,
for example, Tanzanian researchers would not be allowed to travel
to Kenya for scientific meetings because of government policy on
foreign exchange and currency restrictions. In more recent years,
the entire community has been faced with the problem of armed
hostilities between Uganda and Tanzania. In these kinds of situations, EAAFRO officers could have an important role to play in the
future simply because they are regional, community officials who
are more able to cross national borders than are national researchers. One of the earliest identified advantages of the field trials
officers was their ability to cross into Kenya and return with plant
material for the testing program.

    An accepted piece of the conventional wisdom of technology
transfer today is that transfer principally occurs through people--
not through such routes as publications and information systems.
Therefore, a variable called "personal contact transactions," referred to above, was investigated. It was composed of three aspects: the number of EAAFRO officer-visits to the national researcher, at his station, which had occurred in the previous two
years; the total number of visits the national researcher had made
to EAAFRO headquarters; and the number of times national researchers had met EAAFRO officers at scientific conferences,
seminars, or other kinds of meetings. The results to the first two
questions are shown in Tables 6.9 and 6.10. Almost half of the national researchers, 72 out of 165, reported that they had not had a
visit from an EAAFRO officer in the preceding two years. These results do not include visits by EAAFRO field trials officers, but do
include visits from EAAFRO officers in decentralized divisions to
places other than the national stations where they were located.

    This lack of contact occurred primarily in Tanzania, but also
to a significant extent in Kenya and Uganda. Only three Tanzanian
researchers had received more than two EAAFRO officer visits at

their stations in the preceding two years, which was disproportionately lower than the corresponding numbers for Kenya and Uganda.
This tendency may be partly due to the distance factor, the relative
newness of many Tanzanian researchers, travel or monetary restrictions between Kenya and Tanzania, or other related reasons.
In addition, some of EAAFRO officers' visits to Tanzanian research
stations did not result in a favorable experience. More than one
Tanzanian officer remarked about the hurried and uncaring nature
of some EAAFRO visits in which an officer flew in and out in one
morning or stopped by for 10 minutes between game parks.

TABLE 6.9

EAAFRO Officer Visits to National Researchers

Number of National Researchers from	Number of EAAFRO Officer Visits								No Answer	Total
	> 6	6	5	4	3	2	1	0		
Kenya	10	7	3	3	2	9	11	30	--	75
Uganda	2	3	2	4	4	3	7	19	--	44
Tanzania	--	1	--	1	1	5	12	23	3	46
Total	12	11	5	8	7	17	30	72	3	165

TABLE 6.10

National Officer Visits to EAAFRO Headquarters

Number of National Researchers from	Total Number of Visits to EAAFRO Headquarters							No Answer	Total
	> 5	5	4	3	2	1	0		
Kenya	30	1	5	4	9	11	15	--	75
Uganda	2	1	2	2	8	9	20	--	44
Tanzania	--	--	2	2	2	8	29	3	46
Total	32	2	9	8	19	28	64	3	165

In Uganda, the greater number of researchers receiving more than two EAAFRO officer visits may have been due to their relatively greater experience at their stations. In Kenya this may have been due to proximity or the amount of cooperative research, or both, that was taking place. Particularly in the case of Kenyan officers at the National Agricultural Laboratories in Nairobi, it seemed that once a personal contact was made with an EAAFRO officer that many EAAFRO officer visits subsequently occurred--that is, once a relationship was established, it was able to be followed up because the two institutions were so close. Many Kenyan officers at this same station, however, reported no EAAFRO officer visits to them. This may be because that first contact was never established. A minimal amount of visits seem to be made by EAAFRO officers to national stations in all partner states, and if a national researcher reports no such visits to him it may be due to the recency of his arrival or some other explanatory factor. Above this minimal amount, however, the number of EAAFRO officer visits may depend to a large extent on distance and other factors, for example, personal friendships.

From the results in Table 6.9, it is apparent that the largest numbers of national researchers have either not been to EAAFRO headquarters at all, or they have been there many times--too many to remember. It is the Kenyan researchers from the National Agricultural Laboratories in particular that comprise almost the entire latter category. Tanzanian and Ugandan researchers total more than three-fourths of the former category.

Twenty-nine out of 46 Tanzanian researchers had never been to Muguga, and only four had been there more than twice in their lives. Twenty of 44 Ugandan researchers had never been there, but greater numbers had visited three and four times. Even 15 out of 75 Kenyan researchers had never been to Muguga, but this is a small proportion compared to the numbers which had visited a large number of times.

Part of this lack of national travel from Tanzania and Uganda has undoubtedly been due to travel restrictions placed on national researchers by those governments in an effort to keep within tight budgets and conserve scarce foreign exchange. Part is also undoubtedly due to the distance involved. In Kenya, of course, there are no border restrictions for the national researchers and distances to EAAFRO are usually less. Another partial explanatory factor apparently operating in all three partner states was the recency of arrival of some national researchers to their positions. It should also be noted that many of the African national researchers who received their university education in east Africa visited EAAFRO at least once on class tours. The closeness of EAAFRO and Kenya's National Agricultural Laboratories might be noted once again with respect to the relatively high numbers of national researchers from this institution who would casually visit EAAFRO large numbers of times just to use the library or visit socially with friends.

In addition, 46 researchers from Kenya, 30 from Uganda, and 18 from Tanzania reported one or more scientific meetings where they had met EAAFRO officers. In Kenya and Uganda, specialist committee meetings were the most frequently mentioned. In Tanzania, two special conferences the government had recently sponsored on land use and soil fertility, which an EAAFRO representative attended, were mentioned the most. A few national researchers in each partner state mentioned the East Africa Cereals Conference in this context--conferences sponsored by USAID on cereals research in Greater East Africa. There also was some mention in Kenya of seminars sponsored by the National Agricultural Laboratories that EAAFRO officers attended. One meeting forum conspicuous for its absence in the discussion to this question was the East African Academy of Sciences.

Two specific mechanisms related to technology transfer and referred to above will be expanded upon briefly: EAAFRO's decentralized divisions; and specialist committees.

EAAFRO has three decentralized divisions located at different national research stations around the region: the Sorghum and Millet Division, located at Serere Research Station in Uganda; the Maize Genetics Division, located at the Kitale National Agricultural Research Station in Kenya; and the Sugar Cane Breeding Division, which has a breeding center temporarily located at the Coast (Mtwapa) Research Station in Kenya (the permanent site for EAAFRO's sugar cane breeding center is under construction at Kibaha, Tanzania) and a sugar cane disease center located at Kawanda Research Station at Kampala, Uganda. At the time of this study, four EAAFRO officers were stationed at Serere on Sorghum and Millet, one stationed at Kitale on Maize, and one each stationed at Mtwapa, Kibaha, and Kawanda on Sugar Cane.

The question asked of national researchers at those stations where EAAFRO's decentralized divisions were located did not refer to technology transfer, but to "institution-building." Institution-building was defined as the contribution that an individual or unit made to the total institution of which he or it was a part. Part of that contribution could be in the form of physical buildings or equipment, or in the form of research that contributed to the institution's reputation. The concept also included those intangible qualities that contributed to such important institutional attributes as morale, harmony, and spirit. The results are shown in Table 6.11. More research would be needed to draw any conclusions about technology transfer through this mechanism other than what already has been stated.

TABLE 6.11

Perceived Institution-Building Contributions of EAAFRO
by National Researchers at Affected Stations

Responses	Number of National Researchers from				
	Kawanda	Serere	Coast	Kitale	Total
Perceived institution-building					
Very much	--	3	--	6	9
Quite a lot	4	5	1	8	18
Some	5	4	--	1	10
Just a little	6	1	--	--	7
Almost nothing	2	1	1	--	4
No answer	1	--	1	3	5
Total	18	14	3	18	53

Specialist committees were addressed in the study because of their direct connection with EAAFRO's project selection process. That theoretical process is described as follows in EAAFRO's Annual Report for 1969.

> Research requirements are first discussed by the Specialist Research Committees, which may be standing or ad hoc, and which are convened and chaired by EAAFRO specialists. These Specialist Committees are composed of research workers in the appropriate disciplines. . . .
>
> The recommendations of these committees are submitted for approval to the appropriate Research Coordinating Committees. . . .
>
> The recommendations of the Coordinating Committees are duly considered by the Research Council, but the implementaton of the programs approved by the Council is dependent on financial provision being granted by the East African Legislative Assembly.[14]

As envisaged originally by the Frazer Commission, it was the function of the Coordinating Committees to determine research priorities,
and so the point was emphasized that territorial representation was
essential: ". . . if coordination is to be fully successful, it is essential that the Territories should be represented by senior technical
officers having authority to implement agreements which the Committee may arrive at. . . ."[15] Today, it is still pointed out by
EAAFRO that the membership of these Coordinating Committees includes ". . . the appropriate Directors or Commissioner of Agriculture and Veterinary Services, the Chief Conservators of Forests
and representatives of the wildlife organizations of the three National
Governments."[16] There are only four coordinating committees in
fields that are connected with EAAFRO interests: agriculture, animal industry, forestry, and wildlife. Therefore, particularly for
the Agricultural Coordinating Committee, which encompasses a
wide variety of subjects, in order for the Coordinating Committees
to be able to function properly, the Frazer Commission deemed it
necessary that they should have the power to appoint Specialist Sub-
committees with the following terms of reference:

- to report upon the scientific and technical issues involved
in problems submitted to it either by the Coordinating Committee or
by its own members; and
- to recommend problems for inclusion in territorial or
inter-territorial research programmes.[17]

In 1969, EAAFRO scientists were responsible for convening and
chairing the following specialist research committees: agricultural
botany, agricultural meteorology, agricultural machinery, pasture
research, rangeland research, herbicides, entomology and insecti-
cides, coffee, sugar cane, soil fertility, forestry, wildlife, statistics, and plant imports and exports.[18]
    It was the intention of the Frazer Report that the "coordinating
machinery" described above keep the regional research institutes in
touch with the needs of the three territories. This machinery, however, has functioned at varying rates of effectiveness. The starting
point of the project selection process is the specialist committees.
They have met at different frequencies, at different sites, with different agendas, for different lengths of time, and with different
memberships. There were also different philosophical perceptions
of their objectives. Some have functioned as active and searching
bodies, including many different elements from national organizations in order to arrive at a broad consensus of what research should
be undertaken in a certain field. Others have functioned as narrowly

scientific groups--closed meetings of scientific specialists reading
strictly scientific papers and producing a scientific report. In cases
where the specialist committee has been active in recommending research priorities based on national discussions, the coordinating
committee also appears either to be actively functioning or going
along with the specialist committee's recommendations. Otherwise,
the coordinating committees seem to have become inactive.

In investigating the roles of these specialist committees from
the perspective of the national researchers, the only ones where significant project selection decision making and research coordination
was perceived to be taking place were in the fields of plant imports
and exports (plant quarantine), forestry research, and sugar cane.
National researchers' opinions on the effectiveness of other specialist committees were found at the same time to be overwhelmingly
positive, but they defined "effectiveness" in terms of the communication and transfer of scientific information that occurred at these
meetings. Therefore, even though this mechanism was not generally successful in terms of its official objective, it did turn out to
be an appreciated vehicle for technology transfer.

There are more results, barriers, and mechanisms related to
technology transfer in the research study described in this chapter,
but these would appear to be the most important points.

NOTES

1. T. W. Schlie, "Some Aspects of Regional-National Scientific Relationships in East Africa" (Ph.D. diss., Northwestern
University, 1973).

2. F. Kahnert et al., Economic Integration Among Developing
Countries (Paris: OECD Publications, 1969), p. 9.

3. Peter Robson, Economic Integration in Africa (London:
George Allen and Unwin Ltd., 1968), p. 66.

4. Report of the Commission on International Development,
Lester B. Pearson, chairman, Partners in Development (New York:
Praeger, 1969), p. 277.

5. United Nations, Advisory Committee on the Application of
Science and Technology to Development, Third Report to the Economic and Social Council (E/4178), May 1966, p. 17.

6. The Pearson Commission described the EAC as ". . .
perhaps the most important cooperative arrangement involving a
common market and an impressive range of services which are
operated jointly by the three countries," op. cit., p. 95.

7. B. A. Keen, "The East African Agricultural and Forestry
Research Organization--Its Origin and Objectives" (Nairobi: East
African Standard Limited, no date), pp. 7-8.

    8. E. B. Worthington, A Survey of Research and Scientific
Services in East Africa, 1947-1956 (Nairobi: East African High Commission, 1952), p. 14. Worthington also stated the following about
the territorial role in scientific research: "Nevertheless most territorial departments of agriculture, veterinary services, medical
services, etc. find it necessary to maintain groups of specialist officers for routine work and for investigations of a short-term kind on
problems with which the department is directly concerned." Ibid.,
p. 14.
    9. A. C. Frazer, Chairman, Report of the Commission on
the Most Suitable Structure for the Management, Direction, and
Financing of Research on an East African Basis (Nairobi: Government Printer, 1961), p. 25.
    10. East African Academy, Research Services in East Africa
(Nairobi: East African Publishing House, 1966), p. 45.
    11. East African Agricultural and Forestry Research Organiza-
tion, Record of Research--Annual Report 1969 (Nairobi: East African
Community, 1970), p. 2.
    12. Schlie, op. cit., pp. 331-32.
    13. Ibid., pp. 338-39.
    14. EAAFRO, op. cit., pp. 2-3.
    15. Frazer, op. cit., p. 51.
    16. EAAFRO, op. cit., p. 3.
    17. Frazer, op. cit., p. 52.
    18. EAAFRO, op. cit., p. 8.

7

INTERNATIONAL TRANSFER OF OIL TECHNOLOGY AND THE POLITICAL ECONOMY OF THE NIGERIAN OIL INDUSTRY

INTRODUCTION

Despite the multinational character of the world oil industry, excluding the Socialist countries, oil technology still resides exclusively in the advanced countries as proprietary know-how of oil companies. Benefits to oil-exporting countries in the form of inflows of technological know-how have not been evident, and financial benefits from exploration and exports of crude oil have been generally low. Major changes resulting in substantial financial benefits only began to accrue to the exporting countries in the 1960s.

The basic proposition in this chapter is that the absence of tangible long-term technological benefits from upstream and downstream operations in the oil industry, and the lack of penetrative linkages between the activities of the industry and the rest of the economy have significantly limited the overall economic impact of the surge in financial benefits accruing from expansion in oil production and exports.

Typically, upstream operations consist of exploration, discovery, drilling, storage, and some aspects of transporting crude oil, and downstream operations include some aspects of crude oil delivery to refineries, oil refining, marketing, and all other activities related to the sale of oil and oil products for final consumption. The industry's technological impact has been constrained, notwithstanding significant improvements in tax arrangements, and increasing control over the output and the price of oil by the Nigerian government.

A great deal has been written on the history and the fiscal importance of the Nigerian oil industry.[1] However, there is not yet a thorough analysis of the industry's technological importance and its potential linkage effects with the other sectors of the economy. One exception is Scott Pearson's study "Petroleum and the Nigerian Economy," but the industry's technological linkages and contributions were not the central subjects of inquiry. [2]

Although the primary focus of this study is an inquiry into the nature of technology "transfer" to the Nigerian oil industry, two other related aspects of the industry are also examined. These are, the development and growth of the industry, and the economic impact of the financial benefits from export of crude oil. In the examination of technology "transfer" to the oil industry, the long-term technological benefits to Nigeria, and Nigeria's oil policies regarding the acquisition of foreign technology for the development of oil-based industries are also studied. The concluding section is an assessment of the strengths and weaknesses of Nigeria's bargaining position in the acquisition of suitable foreign technology for the active participation of the country's indigenes in the development of oil and allied industries.

The political and economic realities of the international oil industry as they pertain to the Nigerian situation clearly suggest that the emphasis is only on short-term financial benefits. The potential technological benefits that are capable of effecting the industry's permanent linkages with the rest of the economy have been generally neglected, and this could be explained by two factors: the dominance of foreign investment and its direction singularly toward the exploration of crude oil for export to Western Europe and the United States, and the relatively small industrial base that impelled the importation of all capital equipment and the large pool of high-level scientific and technical labor input requirements of the industry.

Unfortunately, these conditions and the policies that led to them have yet to be amply corrected either by the country itself or through assistance by collective action of the Organization of Petroleum Exporting Countries (OPEC).* However, before Nigeria's membership in OPEC in 1971, the mere existence of the organization and the "most favored nation clause" made possible some improvements in Nigeria's position in terms of its bargaining with the foreign oil companies.†

---

*The present membership of OPEC includes Iraq, Iran, Kuwait, Saudi Arabia, Abu Dhabi, Libya, Algeria, Indonesia, Venezuela, Nigeria, Qatar, Ecuador, and Gabon. These countries produce about 80 percent of total world oil exports.

†The provision of this clause is essentially an avowal by the oil companies to accord Nigeria the same terms as their most favorable terms to other governments in the Middle East and elsewhere in Africa, specifically Libya and Algeria.

THE DEVELOPMENT AND GROWTH
OF THE OIL INDUSTRY

Periodic efforts have been made in the exploration for oil in
Nigeria since the early 1900s. The development of the oil industry
began under British colonial rule. Although the Nigerian Bitumen
Corporation, a German firm, pioneered oil drilling activities in
Nigeria in 1908, the British colonial government granted exclusive
mineral oil concession to Shell D'Arcy Petroleum Company in 1938.
The rights of this concession pertained to the entire country. Shell
D'Arcy later became Shell-BP Petroleum Development Co. (Nigeria)
Ltd., a jointly owned subsidiary of British Petroleum and Royal
Dutch Shell, with 90 percent majority ownership divided between the
British government and Royal Dutch Shell (51 percent) and British
private capital (39 percent).

The turning point in the historical importance of geological
surveys and oil exploration occurred in 1956 when Shell-BP made
the first discovery of commercial volume at Oloibiri. Two years
later Shell-BP began exporting crude oil. As Table 7.2 shows, between 1958 and 1965 all of Nigeria's crude oil production was exported abroad. About 80 percent of the oil exports went to Western
Europe. This pattern has remained relatively unchanged. In more
recent years, the United States, France, and the United Kingdom account for about 58 percent of the daily oil exports. Western European countries represent the world's major negative net producers
of oil, and the United States, as well, may be so classified since
1971. (See Tables 7.1 and 7.3.) However, the future dependence of
the United States and Great Britain on oil exports may be considerably diminished by the Alaskan and the North Sea oil discoveries
respectively.

TABLE 7.1

Production and Consumption of Oil in 1971
(millions of tons)

Country/Region	1 Production	2 Consumption	3 Net Production (1-2)
United States	473.0	715.0	-242.0
Western Europe	22.4	652.0	-629.6
USSR	372.0	350.0	22.0
Middle East	803.0	54.0	749.0
Nigeria	75.0	2.4	72.6

Note: 1 barrel = .135 metric ton.

TABLE 7.2

Nigerian Petroleum Production and Exports, 1958-74
(in £N millions)
volume = 1,000 barrels per day

Year	Crude Oil Production Volume	Value	Exports Volume	Value	Local Sales Volume	Value
1958	5	1	5	1	--	--
1959	11	3	11	3	--	--
1960	17	4	17	4	--	--
1961	46	11	46	11	--	--
1962	68	17	68	17	--	--
1963	76	20	76	20	--	--
1964	120	32	120	32	--	--
1965	270	69	266	68	4	1
1966	415	100	383	92	32	8
1967	317	77	300	72	17	5

Source: Compiled from the Annual Report of the Petroleum
Division of the Federal Ministry of Mines and Power, 1965-66,
Lagos, Nigeria, 1967; and Scott Pearson, "Nigerian Petroleum:
Implications for Medium-Term Planning," in Growth and Development of the Nigerian Economy, ed. Carl K. Eicher and Carl Liedholm
(Michigan: Michigan State University Press, 1970), p. 357.

TABLE 7.3

A Summary of Production and Exports Figures, 1970-74

Year	Volume of Production (thousands of barrels per day)	Export Volume (thousands of barrels per day)	Export Value (₦ million)
1970	1,100	1,030	510
1971	1,500	1,434	953
1972	1,800	1,717	1,176
1973	2,100	1,866	1,842
1974	2,400	--	2,842*

*Value of oil export for the first six months of 1974.

Note: Export volume figures represent daily average for the
years 1970-73. The figure for 1974 is not available. For conversion
purposes, £N1 = ₦ 2.00 and ₦1.00 = $1.66 = £0.6836 sterling.

Source: Compiled from various issues of Standard Bank Review, Standard Bank Ltd., London, 1973-74.

TABLE 7.4

Destinations of Nigerian Crude Oil Exports, 1970–71
(in barrels per day)

Export Destination	First Half 1971	First Half 1970
United States[a]	288,922	156,210
France	283,434	73,380
United Kingdom	273,279	220,233
Netherlands	207,429	148,999
Germany (West)	74,202	32,046
Italy	70,339	10,101
Spain[b]	49,603	57,915
Brazil	36,562	33,533
Denmark	33,532	2,649
Sweden	32,715	34,079
Norway	28,463	14,489
Canada	28,146	30,576
Japan	8,083	--
Ivory Coast	7,741	--
Others	38,029	104,290
Total	1,460,479	918,500

[a]Includes 54,103 barrels per day (bpd) primarily to Virgin Islands in first half 1971; 47,055 bpd second half 1970, and 37,303 bpd first half 1970.
[b]Includes 1,374 bpd to Canary Islands in first half 1971; 24,004 bpd second half 1970; 22,994 bpd first half 1970.

Source: Petroleum Intelligence Weekly, October 25, 1971.

During the early phase of the Nigerian oil industry's development, the British colonial government laid down legislation to govern the operations of oil companies. When compared with the terms governing the operations of the same oil companies in Venezuela and the Middle East, the terms of the legislation, in particular, the Mineral Oil Ordinance and the Petroleum Profits Tax Ordinance enacted in 1959, were highly unfavorable to Nigeria's public finance. Four provisions of the ordinances were especially favorable to the oil companies:

- Variable royalties of between 8 and 12.5 percent were assessed on the value of the crude at the extraction point.*
- A 50 percent share of profits due to government also included royalties and other duties.
- The reference price for the assessment of profits for tax purposes was the "realized price."
- The depreciation rates allowed the oil companies were excessive and had no rational economic basis. The first year rate was usually in excess of 50 percent of the expenditure on plant and equipment.[3]

These provisions and other arrangements between the oil industry, which was dominated singularly by Shell-BP, and the Nigerian government, precluded bilateral agreement on prices, and the ability of the government to obtain equitable prices for oil was also restricted. These conditions and the absence of a stable basis for estimating government revenues from oil contributed to the persistent shortfall in revenues throughout the 1960s. Similar shortfalls represent one of the major problems in the implementation of economic development planning in virtually all DCs that are dependent on one or two sources of foreign exchange.

Since the late 1950s, other oil companies, namely Mobil Oil, Gulf, Tenneco, Agip, Safrap, Phillips, Texaco/Chevron, and Esso, have penetrated the industry, but Shell-BP continued to enjoy substantial monopoly advantage until Nigeria's national political independence in 1960. Since the late 1960s, there have been more than a dozen companies prospecting for oil in Nigeria, but the industry is oligopolistic in character.

Notwithstanding this structure, Shell-BP still has considerable advantage. It remains the largest of all the companies in terms of its share of the industry's production and crude oil exports. Figure 7.1 shows the dominance of Shell-BP in the distribution of oil prospecting licenses and mining leases and the location of concessions in the Niger Delta. Before the establishment of the Nigerian National Oil Corporation (NNOC) in 1971, Shell-BP maintained control over the production of crude oil as it had since the inception of the industry. In 1972 and 1973, it was responsible for as much as 66 percent of all crude oil production. Although its percentage share of production has declined, it is still the dominant company in the industry.

Rather than seek control, revise the unfavorable colonial terms identified above, and plan for long-term growth, upon independence,

---

*This is in comparison to royalty payments of 16.7 percent on posted price in Venezuela, for example.

FIGURE 7.1

Niger Delta Concessions
(oil prospecting licenses and mining leases)

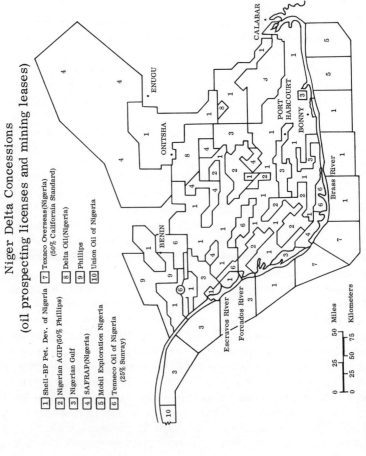

Source: <u>Petroleum Press Service</u>, March 1970, p. 86.

140

Nigerian politicians' attention was focused on regional rivalries and ethnic competition for a larger share of national oil wealth and political power. Inevitably, oil revenues and the promise of greater wealth to come from the industry figured prominently in the rivalry that precipitated the 1967-70 civil war.

The growth in oil production as shown in Table 7.2 was gradual throughout the pre-civil war period. It rose from 5,100 barrels per day (bpd) in 1958 to about 500,000 bpd in 1966 before the outbreak of the civil war. Crude oil production declined during the civil war, but has risen considerably since the end of the war. Major policy changes on production began to take effect in late 1969. In 1970, production climbed to 1.1 million bpd and thus put Nigeria among the top ten oil-producing countries, excluding the Socialist countries. In 1971, production rose to 1.5 million bpd; in 1972, 1.8 million bpd; in 1973, 2.1 million bpd; and in 1974, about 2.4 million bpd. Table 7.6 shows the production distribution among the major oil companies for 1973 and 1974.

The spectacular growth since 1970 has put Nigeria seventh in ranking among the world's oil-producing countries, outside the Socialist bloc, and the major oil exporter to the United States in 1974 excluding Canada. See Table 7.5. The trend in the growth of petroleum and natural gas is illustrated by the index of mineral production in Table 7.7 with comparative information on other mineral resources. See also Table 7.8 for the estimated reserves of petroleum and natural gas. The index of mineral production, of which crude petroleum accounts for about 88 percent of the aggregate weight index, has shown the strongest increases when compared to any other single sector of the economy. Based on the official planned estimates of 10 percent each year in the minimum average rate of growth of crude oil production, by the end of the Third National Development plan in 1980, crude oil production would be about 4.6 million bpd. Although this is a feasible target given comparative data on proven and probable reserves shown in Table 7.9, a stable price at about $12.00 pb and a production ceiling at the 1974 level of 2.4 million bpd would leave ample room for the fulfillment of the projected investment requirements of the third plan estimated at about ₦30 billion, or approximately $49 billion.

The effectiveness of policy changes on expansion in production and the promotion of improvements in revenue through new legislation are being manifested in the general posture of the government in relation to the new oil concessionaires and the periodic renegotiation of contracts with existing oil companies for equitable rearrangements in the distribution of the financial benefits from the exploitation of Nigerian oil.

TABLE 7.5

Selected U.S. Imports of Crude Petroleum and Refined Products
(thousands of barrels and thousands of dollars)

	Total Crude	Percentage of Crude Imports	Total Imports	Percentage of Total Imports	Value	Percentage of Value
Imports from OPEC						
Algeria	75,899	5.55	82,736	3.76	1,083,359	4.44
Ecuador	23,599	1.73	23,697	1.08	268,459	1.10
Indonesia	103,822	7.59	110,795	5.03	1,271,700	5.21
Iran	186,572	13.65	189,490	8.60	2,017,974	8.27
Iraq	--	--	--	--	--	--
Kuwait	576	.04	576	.03	5,928	.02
Libya	--	--	*	*	3	*
Nigeria	247,501	18.10	256,013	11.62	3,223,644	13.21
Qatar	6,666	.49	6,825	.31	78,702	.32
Saudi Arabia	150,260	10.99	158,112	7.18	1,662,156	6.81
United Arab Emirates	31,731	2.32	31,834	1.45	364,567	1.49
Venezuela	155,728	11.39	429,886	19.51	4,365,734	17.90
Total	982,354	71.8	1,289,964	58.5	14,342,226	58.7
Major exporters, non-OPEC						
Canada	311,143	22.7	395,249	17.94	3,901,478	15.99
Trinidad	29,254	2.16	102,370	4.64	1,222,416	5.01
Netherlands Antilles	--	--	184,496	8.37	1,934,681	7.93
Total selected imports	1,327,569	97.0	1,981,354	89.9	21,511,036	88.2
Total imports of petroleum and petroleum products	1,367,322	100.0	2,203,391	100.0	24,395,002	100.0

*Less than 1,000 barrels or .01 percent.

Note: Does not include imports from or into Puerto Rico, Guam, or the Virgin Islands.

Source: Federal Energy Administration, National Energy Information Center, mimeographed, February 28, 1975.

TABLE 7.6

Nigerian Oil Production by Company
(millions of barrels per day)

Company	October 1973	October 1974	Percent Change
Shell-BP	1,373	1,438	4.7
Gulf	383	392	2.3
Mobil	246	247	0.4
Agip/Phillips	122	166	36.0
Elf	63	82	30.1
Texaco	8	n.a.	n.a.
Total	2,195	2,325	5.9

n.a. = not available.

Note: The period covered coincides with the period of the energy crisis, and Nigerian crude oil production as well as export grew during the entire period. In 1974 Nigeria was the major exporter of crude oil to the United States, accounting for about 25 percent of U.S. oil imports.

Source: The Standard Bank Ltd., Standard Bank Review, London, December 1974, p. 13.

TABLE 7.7

Index of Mineral Production
(base: quarterly average 1965 = 100)

	Weight	1966	1967	1968	1969	1970	1971[b]	1972[c]
Cassiterite (tin ore)	36.7	97.5	97.9	101.1	90.3	82.1	75.6	71.8
Coal	7.8	86.3	13.0	--	1.8	8.2	26.0	45.5
Columbite	3.9	80.8	70.0	41.3	54.3	57.9	49.5	47.0
Gas (natural gas)[a]	1.9	182.8	179.7	152.4	66.2	115.1	192.5	253.5
Limestone	3.9	85.1	64.6	50.1	54.2	52.7	61.6	103.9
Petroleum	381.6	152.3	116.4	50.6	197.1	389.1	568.5	662.5
Total	435.8	145.4	112.4	54.3	181.5	349.3	506.5	589.4

[a]Consumption.
[b]Revised.
[c]Provisional.

Source: Central Bank of Nigeria, Annual Report and Statement of Accounts, December 31, 1972, p. 22.

TABLE 7.8

Estimated Nigerian Reserves of Petroleum and Natural Gas, 1969-73

	Petroleum (millions of barrels)	Natural Gas (billions of cubic feet)
January 1, 1969	4,000	3,900
January 1, 1970	5,000	5,000
January 1, 1971	9,300	6,000
January 1, 1972	11,680	40,000
January 1, 1973	14,000	40,000

Source: The Oil and Gas Journal, December 30, 1968, p. 103; December 29, 1969, p. 95; December 28, 1970, p. 93; December 27, 1971, p. 73; December 25, 1972, p. 83

TABLE 7.9

World Oil Production and Reserves, 1973

Country	Population (millions)	Proved and Probable Billions of Barrels of Oil	Output (millions of barrels per day)	Years of Reserve at 1973 Output Rate	1973 Producing Position
Abu Dhabi	.1	21	1.3	45	12th
Algeria	14.7	47	1.0	20	13th
Canada	22.1	11	1.8	17	10th
China (Peoples Republic)	800.0	n.a.	1.0	n.a.	14th
Indonesia	125.0	12	1.3	22	11th
Iran	31.9	65	5.9	28	4th
Iraq	10.4	29	2.0	44	9th
Kuwait	.9	65	3.0	66	6th
Libya	2.1	30	2.1	32	7th
Nigeria*	73.4	14	2.0	27	8th
Qatar	.2	7	.5	31	15th
Saudi Arabia	8.1	145	7.5	51	3rd
United States	212.0	48	9.2	15	1st
USSR	250.9	n.a.	8.5	n.a.	2nd
Venezuela	11.3	14	3.5	11	5th

*Nigeria's percent of world total reserves = 2.31 percent.

n.a. = not available.

Sources: Compiled from Hollis B. Chenery, "Restructuring the World Economy," Foreign Affairs 52

Among the initial changes in oil legislation are the 1966 Income Tax Amendment Decree, and the 1967 Petroleum Tax Amendment Decree. The resultant improvements in the government's financial position included a reduction by about half in the depreciation rate for capitalized investment by oil companies and, hence, an increase in the profit taxes collected by the government, the establishment of posted instead of listed prices as the basis for tax calculations, and the assessment of royalties as operating expenditures.

A new era in the relations between the oil companies and the government began with these developments. The substantial surge in production since the end of the civil war in January 1970 has been accompanied by more periodic revisions in agreements between the oil companies and the government. Social and political lessons from the war, and the subsequent firm political control by the military government have considerably attenuated past regional and ethnic political rivalry for oil revenue. Attention has been shifted to company-government relations in the search for government control over oil production and greater influence on the price of oil and to major reallocation of the revenue derived from oil.

The major oil revenue reallocation which went into effect in April 1975 is given below:

Oil Revenue Source

Mining Rents and Royalties

	On-shore production	Off-shore production
Old Allocation	1) 45 percent to state where production occurs. 2) 5 percent to federal government. 3) 50 percent to distributable pool.	1) 100 percent to federal government. No allocation to distributable pool.
New Allocation (effective April 1975)	1) 20 percent to state where production occurs. 2) 80 percent to distributable pool. No allocation to federal government.	1) 100 percent to distributable pool. No allocation to federal government.

In 1971, government tax take was revised upward by 5 percent
to 55 percent for 34 gravity crude oil.* This agreement was subsequently revised to take appropriate account of the 1971 dollar devaluation. The revision became effective in February 1972. By early
1973, a number of negotiated price adjustments brought the tax reference price of 34 gravity crude oil up to $3.42 per barrel, and gov-
ernment tax take rose to $1.90 per barrel. Just before the major
rise in the price of oil in October 1973, the posted price of 34 gravity crude was at about $4.29 per barrel. By the end of 1974 the
posted price of Nigerian oil had risen to $14.60.

The Oil Industry and the Foreign Trade Sector

Before the civil war, particularly between 1958 and 1964, imports by the oil industry exceeded the value of crude oil export by as
much as ₦5 million. This adverse effect on the country's balance of
payment reflects the capital intensity that characterized the initial
phase of the industry's development. Although export earnings, government tax take, and spending locally by the oil industry supplied
the needed foreign exchange for imports, the oil industry's imports
have historically accounted for the largest outflow of foreign exchange. In 1965, direct and indirect import cost totalled ₦22 million, while the industry's total contribution to Nigeria's balance of
payments amounted to ₦35 million. See Table 7.2 for export value
of crude oil from 1958 through 1974. In 1966 imports of materials
and services, excluding capital equipment, associated with crude oil
exploration and drilling had risen to ₦58.0 million. Foreign exchange outflows for imports totaled 60 percent of oil companies' expenditure during the 1960s. The percentage of total foreign exchange
earnings from oil retained in the local economy, that is, inflow of
foreign exchange less outflow for imports, profits repatriation, and
other payments abroad by oil companies that constitute leakages from
the spending stream in the local economy, declined from 67 percent
in 1963 to 38 percent in 1969. The decline in the late 1960s took
place in spite of import controls during the civil war, 1967-70. By
1971 the diffused state of the industry that characterized the war
period had begun to subside. While imports declined by about 16

---

*Nigeria's crude oil is mostly high quality, light gravity oil in
Degrees API. It also has higher proportion of gasoline content relative to fuel oil, and is generally low in sulphur content. The last characteristic is considered highly desirable from the standpoint of en-
vironmental concern in importing countries.

percent in 1971, and government tax take increased, and foreign exchange contributions as a percentage of gross output rose from 61
percent in 1970 to 69 percent in 1971. Since 1970, the total contribution of oil companies to the balance of payments has increased
more steadily than previously. See Table 7.10. Oil refining locally
had some favorable effects on the balance of payments by reducing
the annual import bill for petroleum products. Appropriate expansion in refinery capacity should further free foreign exchange to
meet other development project needs.

TABLE 7.10

Contribution of Oil Companies to the
Balance of Payments, 1969-73
(in ₦ million)

	1969	1970	1971	1972	1973*
Payment to government or government authorities	53.8	176.4	542.4	729.1	1,001.0
Other local expenditure	57.8	76.8	113.2	129.6	114.7
Variation in cash holdings	-0.8	4.0	-12.2	-4.3	3.9
Local receipts	-4.2	-24.0	-38.6	-54.5	-33.2
Total contribution to balance of payments	106.6	253.2	604.8	799.9	1,086.4

*Provisional.

Source: Central Bank of Nigeria, Annual Report and Statement
of Accounts, December 31, 1971; 1972 and 1973, Lagos, Nigeria.

The rapid expansion in exploration, production, and exports of
the Nigerian oil industry since 1963 has caused the oil sector's swift
displacement of the agricultural sector as the chief export and major
source of foreign exchange. While oil exports continued to rise
sharply in the postwar years, the value of nonoil exports, agricultural exports in particular, declined by 24.1 percent between 1971
and 1972. In 1971, oil represented 74 percent of total exports. It
rose sharply to 82 percent in 1972. Crude oil clearly dominates the

recent spectacular increases in the value of exports. See Table 7.11. In 1973, oil accounted for 80 percent of total foreign exchange earnings, and crude oil alone accounted for 92 percent of total exports in mid-1974. This reached an impressive ₦2,842.5 million, which is explained to a great extent by the four-fold rise in price from ₦14.8 per metric ton in 1973 to ₦56.8 per metric ton by mid-1974. Based on provisional figures for 1973 and 1974, the value of other traditional export commodities also went up by 45.8 percent to ₦207.4 million in 1974.[4]

The growth of the oil industry and the resultant shifts in the country's import and export structures have failed to contribute significantly to basic structural shifts and viable linkages with other important sectors of the economy. As a consequence of the national policy of import substitution in manufacturing, there have been major increases in capital goods imports. This, in addition to the increasing imports associated with the vigorous growth in the oil industry, has been producing a gradual shift in the country's import structure toward plant and machinery. The rise in imports has had major offsetting effects on the expanding financial benefits generated by the oil industry and on the balance-of-payments account as well. Nonetheless, in strictly financial terms, it has become evident in recent budgetary allocations for proposed development programs that the positive net effect of the sharp rise in oil revenue has been substantial and this trend is expected to continue, given the increasing bargaining strength of OPEC, and the commitment of the Nigerian government to the full mobilization of oil resources for development purposes.

Public Finance and Financial Benefits from Oil

In the relatively short history of the Nigerian oil industry, the increases in oil leases, exploration rights, and the growth in oil production have caused major financial contributions to government revenue, but this was not evidenced until decisive legislative changes in the financial terms of oil began to take effect in the late 1960s. A reversal of the colonial governments' provisions was then set in motion. The modus operandi before 1960 was highly favorable to the oil companies and entailed financial disadvantage to the Nigerian government and the domestic economy. Since 1966, except during the period of the civil war, the oil industry has begun to make positive contributions to Nigeria's balance of payments, and due to new financial arrangements with the oil companies, government revenues-- for example, receipts in the form of premiums, royalties, rental fees, and profit tax--from the industry have been soaring. Oil

TABLE 7.11

Nigeria's Foreign Trade, 1971-74
(in ₦ million)

	1971	1972	1973	1974[a]
Imports				
Food	88.0	95.1	126.3	68.0
Beverages and tobacco	4.4	4.3	5.2	5.0
Crude materials	20.6	20.7	27.0	19.0
Mineral fuels	9.0	10.3	13.5	23.0
Oils and fats	0.7	1.1	1.4	1.0
Chemicals	122.0	102.6	133.4	81.0
Manufactures classified by materials	319.4	267.9	323.9	203.0
Machinery and transport equipment	428.8	392.6	491.6	274.0
Miscellaneous manufactured articles	70.6	83.1	94.2	46.0
Miscellaneous transactions	15.3	9.0	8.5	5.0
Total	1,078.9	986.7	1,224.8	725.0
Exports				
Cocoa beans	143.2	101.2	112.4	107.0
Cocoa products	10.4	12.2	16.0	14.0
Groundnuts	24.4	19.2	45.4	7.0
Groundnut oil and cake	19.6	16.8	23.0	4.0
Palm kernels	26.0	15.6	18.9	11.0
Palm kernel oil	6.2	5.6	10.0	10.0
Palm oil	3.4	0.2	--	--
Rubber	12.4	7.4	19.4	22.0
Raw cotton	11.0	0.6	4.7	--
Timber, logs, sawn and plywood	7.0	8.0	14.0	9.0
Tin metal	24.8	19.2	15.0	10.0
Re-exports	12.6	13.0	9.0	--
Other	77.3	75.9	148.8	--
Total nonoil	378.3	294.9	436.6	194.0[b]
Crude petroleum	953.0	1,156.9	1,842.0	2,842.5
Total	1,331.3	1,451.8	2,278.6	3,036.5[b]

[a]Provisional figures for the first six months.
[b]Excludes re-exports and other exports.

Note: Nigeria adopted a decimal system of currency in January 1973. The unit Naira (₦) = £N 0.5.

Source: Federal Office of Statistics, Lagos.

revenue accounted for only 3 percent of total government revenue in 1963, or 27 percent of the industry's gross output; and in 1967, before the outbreak of the civil war, it accounted for 17 percent of total tax revenue.

The revision in oil agreements in 1971 increased government tax take and provided for other premiums that virtually doubled the effective tax rate on oil exports and substantially increased government revenue to about 60 percent of the industry's gross output. In 1972, oil alone accounted for about 72 percent of government revenue. This had since risen to 75 percent in 1973. As of December 1974, petroleum revenues as a percentage of total government revenues had risen to 92 percent, while the daily revenue from crude oil exports had also risen to ₦18 million, or approximately $29.0 million, and the daily production rate had risen slightly, from 2.1 million barrels per day in 1973 to 2.4 million barrels per day. Although government revenue from oil has increased substantially in the last four years, it started from a very small base because of the rather modest payments by foreign oil companies before 1970.

In general, the Nigerian government's financial benefits from oil have increased relative to total oil company expenditures and as a proportion of current government revenue. During the 1950s and the 1960s, these financial contributions had been considerably less in Nigeria than elsewhere. Between 1957 and 1966, government revenue from oil exploration and production in Nigeria averaged ₦193 per 1,000 barrels compared with ₦313 in Saudi Arabia and ₦375 in Venezuela. (See Tables 7.12 and 7.13.) Unlike elsewhere, government receipts in Nigeria, as depicted in these tables, included lump-sum payments of premiums for exploration rights and therefore, overstated actual government revenues. Further gross discrepancies are evident in earnings from upstream and downstream activities in the world oil industry. For example, in the 1960s, government tax-take in Nigeria amounted to 42 cents per barrel, while the governments of Libya, Algeria, Kuwait, and Venezuela earned between 85 and 97 cents per barrel and the Western European governments earned between $4.00 and $8.00 per barrel from the processing into final and intermediate products of the crude oil imported from producing countries. Nigeria's membership in OPEC and the latter's international standards for the operations of oil companies have greatly facilitated improvement in the terms offered by the major oil companies to the Nigerian government.

Unlike the steady surge in export earnings and the rising government tax take from oil, spending by oil companies in the local economy and its portion going to Nigerians have fluctuated widely. Payments for local goods and services, and wages and salaries as a percent of total annual expenditures by oil companies declined from

TABLE 7.12

Comparative Government Revenues from Oil Exploration and Production, 1956–66
(in £N per 1,000 barrels)

Year	Kuwait	Saudi Arabia	Iran	Iraq	Venezuela	Libya	Nigeria*
1956	273.2	296.1	301.1	319.6	311.4	--	--
1957	284.3	315.0	310.0	332.5	367.9	--	61.2
1958	291.8	291.8	317.9	317.5	398.6	--	35.0
1959	277.9	270.7	298.6	294.3	351.4	--	334.1
1960	272.9	267.9	286.1	280.7	318.6	--	183.8
1961	265.7	269.6	270.7	273.2	331.8	223.9	407.6
1962	267.1	273.2	266.1	273.9	347.1	231.1	338.1
1963	265.4	281.1	284.6	288.2	352.1	232.1	169.6
1964	262.5	292.9	292.1	286.1	340.7	224.6	147.2
1965	268.2	296.4	296.1	291.8	339.3	299.3	130.4
1966	269.3	308.6	297.1	290.4	325.7	310.7	123.6

*The figures for Nigeria correspond to fiscal years from April 1 to March 31.

Note: For conversion £N1 = $2.8 = ₦2.00.

Source: L. H. Schatzl, Petroleum in Nigeria (Ibadan, Nigeria: Oxford University Press, 1969), pp. 172–73.

TABLE 7.13

Comparative Government Revenues from Oil Exploration and Production, 1956–66
(in £N millions)

Year	Kuwait	Saudi Arabia	Iran	Iraq	Venezuela	Libya	Nigeria*
1956	110.7	107.1	54.6	68.9	263.6	--	--
1957	120.7	115.4	76.1	48.9	345.7	--	--
1958	151.8	110.7	88.2	80.0	354.6	--	0.1
1959	144.6	112.5	93.9	86.8	330.7	--	1.7
1960	166.1	126.8	101.8	95.0	313.2	--	1.2
1961	166.1	142.9	107.5	95.0	335.0	1.1	8.5
1962	187.9	161.1	119.3	95.4	382.5	13.9	8.5
1963	198.9	179.3	142.1	116.1	395.0	38.9	5.0
1964	223.6	200.4	169.6	126.1	400.7	70.4	8.1
1965	228.2	233.2	190.7	133.9	401.8	132.5	14.6
1966	242.9	287.5	216.8	140.7	378.2	170.0	19.0

*The figures for Nigeria correspond to fiscal years from April 1 to March 31.

Sources: Petroleum Press Service 34, no. 7 (July 1967): 246ff.; L. H. Schatzl, Petroleum in Nigeria (Ibadan, Nigeria: Oxford University Press, 1969), pp. 172–73.

44 percent in 1963 to 30 percent in 1966, partly as a result of the disruption caused by the disturbances that eventually led to the civil war. It further declined during the war and reached 11 percent in 1971. The past fluctuations and this eventual decline may also be explained by the consistent pattern of disproportionately large factor payments to expatriate manpower. In 1971, less than 30 percent of the total annual expenditures by the oil companies went to indigenous entrepreneurs. A substantial part of the remaining 70 percent went to expatriates, mostly contractors, geophysicists, and geologists.

The sum of all financial resources from oil, such as government oil revenues and spending in the local economy, are being absorbed without difficulty for the public finance of essential services as well as sectoral and national economic development projects. If there is any one major implication of the steady rise in oil revenue since the late 1960s, it is the considerable mitigation of the persistent short-fall in government revenues that has forestalled the implementation of the planned structural transformation of the economy during the 1960s.* Unlike the Middle East countries, where substantial surpluses are being accumulated, Nigeria does not face the limited scope for internal investment that is currently the case in many of these countries. Therefore, at best, some short-fall in revenue will continue to exist, more so in light of the demand for loans by the International Monetary Fund from the Nigerian government to aid the Third World countries adversely affected by the soaring prices of oil during the 1973-75 energy crisis; and the expensive development projects carried over from the Second National Development Plan 1970-74 and those in the Third Plan 1975-80.[5]

THE STRUCTURE OF THE OIL INDUSTRY

The oil industry belongs in a special category in the Nigerian industrial sector for three primary reasons: the relatively capital-intensive requirements of oil exploration, drilling, and production including refining; the dominance of foreign investment and invariably the complete foreign control of the industry before 1971; and its potential role in the flow of technological know-how and the establishment of linkages with related sectors of the economy such as petrochemical and agro-allied industries.

---

*The short-fall also is demonstrated by a sharp decline in the revenues of foreign exchange from £223 million in 1959 to about £48 million in 1968.

The intensity of physical capital investment and the dominance of foreign investment, and entrepreneurship, encompass the two important modes of transferring technological know-how; namely, machines and humans. These two aspects in the structure of the industry are first discussed before an examination of the industry's technological contributions to the economy and its linkages to key sectors.

Capital Intensity of Oil Exploration, Drilling, and Production

Input combinations in most of the phases of the oil industry require large financial resources, highly sophisticated equipment, and high-level technological manpower, for example, geologists, geophysicists, and engineers. The deficiency in the domestic supply of these inputs invariably left out Nigeria's indigenes from the mainstream of the oil industry. Furthermore, the absence of long-range planning of the industry has also forestalled its full sectoral integration with the rest of the economy.

Although the industry is relatively physical capital-intensive, the staggering cost figures typically used to illustrate the enormous investment requirements for oil exploration and production are attributable more to the riskiness of oil exploration, accounting for dry as well as productive oil wells, than for physical capital investment. In the 1960s, the average cost of finding and developing one barrel of crude oil, excluding production, was $1.45 in the United States, four cents in the Middle East, and fifteen cents in Africa.[6] Major progress in drilling techniques since the 1960s is expected to lead to significant improvement in the conditions of exploration and lower average cost in the future.

The high-fixed costs characteristic of investment in the early phases of the activities in the industry, namely exploration and drilling, and the financial requirement of the "oil lifting" phase usually compel rapid attainment of full capacity rate of production for full advantage of economies of scale by the subsidiaries of the oil companies. However, the consideration of global supply and demand factors by the parent companies tends to discourage the rapid attainment of full capacity rate of production.

The physical capital intensity associated with the early phase of the industry's development has engendered extensive, favorable tax benefits in the form of accelerated capital consumption allowances, and substantial repatriation of profits for oil companies. Significant fiscal measures by the Nigerian government, including tax offsets, were used liberally to favor the oil companies. This policy continued throughout the 1960s. The specific measures used attracted foreign private capital investment and the associated technological know-how. The inflow of foreign investment exclusively through the oil companies that constituted an oligopoly in Nigeria, propagated enclave in the industry, and decisions on items like process design, plant design, site selection, and equipment selection, are made exclusively by these foreign oil companies. Furthermore, attendant technological know-how is commonly treated as proprietary and presumed by government policy makers to preclude beneficial acquisition and diffusion.

One probable explanation for this outlook is the disaccord between oil companies' investment objectives and long-term technological development for the implementation of domestic industrial development programs. A notable exception is the industry's technical training programs, which are intended to produce highly skilled Nigerians. This aspect of oil companies' contributions is examined in some detail below.

The industry as a matter of common practice also imports most of its intermediate goods and services. Consequently, backward and forward linkages with economic activity undertaken by indigenous entrepreneurs are minimal, and the spread effects of investment undertaken in upstream operations are also limited. However, the significant extent of the earnings from oil could more than compensate for the weak linkages through equitable distribution of earnings from oil and the maximization of its impact on the domestic economy. Forward linkages into processing, exporting, and marketing of crude oil, finished oil products and natural gas, petrochemical and agro-allied activities, are generally both physical- and financial-capital intensive, and before 1971 were also heavily undertaken by direct, private, foreign investment--DFI.

Private Investment and Entrepreneurship

Unlike light industries--sawmilling, bread-making, furniture-- in which Nigerian private capital investment and entrepreneurship participated in their development, the oil industry has been exclusively a foreign operation. DFI in oil has risen from approximately one-third of total DFI in the 1950s, to about 60 percent, or $380 million, in 1970. When all the extractive industries are combined, they account for between 70 and 85 percent of total DFI in Nigeria. Low-savings ratio and difficult access to the international capital markets account for the preclusion of participation by domestic investors and entrepreneurs in the early development of the oil industry. Before recent major government equity participation and the establishment of the Nigerian National Oil Corporation (NNOC) in 1971, the exploration, production, export, and marketing of crude oil and all petroleum products and derivatives thereof, were undertaken by foreign oil companies and their subsidiaries or affiliates in Nigeria and abroad. The vertically integrated character of the oil companies impeded free entry of indigenous entrepreneurs into the oil-related service industries. * The prevailing vertically integrated structure of the firms has constrained the industry's capability of inducing investment in other activities, especially for backward linkages. The oil companies sought and, until recently, maintained control over sources of crude oil in order to produce it rather than buy, and also safeguard raw material supplies for their refining and marketing affiliates. This meant that the oil companies were capable of predesignating foreign outlets for the processing of oil. One of the implications of the continuing high volume crude oil exports for processing into highly priced chemical products has been the relatively low value-added in the producer countries. Opportunity for the development of indigenous service industry for oil-related activities like drilling, tubes and pipes fabrication and fitting, cement, and engineering design also has been impeded by the vertical integration in the industry.

In addition to relying on a number of other foreign firms/ contractors for the supply of services, the industry also imports most of its nonlabor input requirements. Although the heavy reliance on foreign sources for primary and intermediate inputs--goods and services--has been explained by the absence of capital goods producing industries and the shortage of appropriate scientific and technological manpower in Nigeria, these foreign sources are often affiliates of the parent companies.

Payments for services of expatriate firms have been increasing, while payments to indigenous entrepreneurs in construction and marketing remain low and the industry's impact in stimulating new major indigenous business ventures also remains minimal. The few indigenous entrepreneurs are generally concentrated in areas like pipeline installation and other labor-intensive service activities.

----

*Unlike the virtually complete vertical integration of major oil companies, particularly the U.S. companies, in their operations in the producer countries, the integration of their structure and operations in the U.S. market is incomplete, especially in downstream operations where the benefits of economies of scale and the considerable extent of market-enabled small independent oil marketing companies to buy from the "majors" usually at the refinery stage of downstream operations and sell to the final consumer under off-brand names.

About 33 percent of oil companies spending locally go to Nigerian entrepreneurs in the construction and service industries. Like all the other spheres of the industry's activities, foreign entrepreneurs and contractors accounted for about 70 percent of the industry's expenditure outlay of ₦72 million in 1965. Wages and salaries to indigenous employees amounted to about ₦1.5 million. The same condition prevailed until 1971.

In 1970, out of the ₦65 million spent by oil companies for services, only 10 percent went to indigenous firms. The rest constitute payments to foreign firms and are subject to leakages out of the Nigerian economy. In general, before major changes in government policy toward the industry in 1971, the total expenditures by oil companies, not subject to leakages, have been estimated to be less than 15 percent of the annual turnover of their operations in Nigeria. Whether the Nigerian oil industry would have existed and developed in the manner it has is moot. What is important is that evidence indicates the existence of major discrepancies in the distribution of the benefits from oil, which has been to Nigeria's disadvantage, especially before the late 1960s.

The factual extent to which the economic dominance of foreign capital, the repatriation of profits and other forms of leakages have offset the positive benefits from the DFI inflow itself is an important empirical question that is difficult to answer conclusively because of the secrecy and a lack of access to information that is necessary for such a verification. Nonetheless, on the basis of the diffused data available, one could safely surmise that the dominance of DFI in all the phases of the industry's activity has effectively impeded the participation of indigenous entrepreneurs, and likewise conferred undue advantage on the oil companies in the distribution of earnings from oil. Since the late 1960s, Nigerian entrepreneurs have begun to develop the capability to provide most of the services currently undertaken by expatriate firms. These include catering services, transportation, pipe laying, and the supply of manpower for drilling activities. Given vertical disintegration and a relatively free market in the transfer and acquisition of oil technology, the requirements for meeting a broader scope of participation by indigenous entrepreneurs in the service and other oil-related industries can be met by minor improvements upon existing modes of transfer and acquisition of know-how, existing manpower and government legislation for their effective use. How to promote these conditions, and meet the requirements for indigenous participation, are vital to the future role of the NNOC and the Nigerian oil industry.

THE TRANSFER OF TECHNOLOGICAL KNOW-HOW

There are two principal media through which oil technology is presumed to have been transferred to the Nigerian oil industry. These are direct private foreign investment in plant and equipment, and the inflow of foreign technical personnel or skilled labor, their interactions with indigenous personnel, as well as the technical training of Nigerians locally and abroad by oil companies. However, the extent of actual technological contributions by the oil industry to the Nigerian economy has been incidental and minimal. Several factors account for this condition. These include: (1) the primary interest of oil companies in Nigeria, in crude oil extraction and its export to Western Europe and North America; (2) the conduct of research and development exclusively outside Nigeria--a practice not unique to the oil industry;[7] (3) heavy reliance on expatriate personnel usually from the country represented by the parent company; and (4) the vertically integrated structure of oil companies. This typically embraces all phases of the industry from crude oil exploration to the marketing of oil products.

The existence of these factors in the development of the industry has effectively limited active participation by Nigerians in upstream or downstream activities or both and thereby hampered the direct and indirect acquisition of appropriate technical know-how through the two principal media just described.

The last of the four factors listed above is perhaps the most crucial because the extent to which the practices associated with the other three factors are maintained depend on the completeness of the vertical integration of the industry. The implication of the third factor is low contribution to the industry's value-added by indigenous factor inputs measured in terms of wages and salaries and expenditure of the government less intermediate goods. In addition to these two payments, Scott Pearson included in his study harbor dues and port charges in measuring the industry's value-added to gross national product. It ranged from £N7. 7 million in 1963 to £N32. 0 million in 1967, with an average growth rate of 45. 8 percent during the four years.[8] The structure prevalent in the industry, particularly the fully developed integration forward into marketing by major oil companies, poses an apparent barrier to the entry of indigenous entrepreneurs and their active participation in downstream operations, thus limiting "competition" to the few major oil companies constituting the international oil cartel. *

---

*The importance of oil and the sharp rise in the bargaining strength of OPEC since 1970 has meant the virtual supercession of the international oil companies' cartel by producer countries'

It is understandable why the oil companies desire to maintain exclusive rights to their technological know-how. The question--how could the international oil companies enable producer nations to master the technological know-how of the oil industry?--has been asked repeatedly by the producer nations, but for obvious reasons no definitive answer has been forthcoming from the oil companies. Clearly, it is this know-how and the expertise to use it effectively that ensures the indispensability of the oil companies to the producer countries. Thus, the game theoretic assessment of the posture taken by oil companies vis-a-vis the producer countries suggests a zero-sum-game situation. On the one hand, making the technological know-how available on a nonproprietary basis to the producer countries would deprive the oil companies of their advantages and the useful basis of their existence in these countries; on the other hand, the issue of a fair exchange between both parties remains perennially unsettled. The expectation that once the producing countries have acquired appropriate competence in oil technology they would seek the complete control of their respective oil industries has also posed a potential, but major competitive threat to the vertically integrated structure prevalent in the industry, and has thus forestalled easy availability of oil technology to producer countries. There is an apparent concern among major oil companies over the prospect that even if technological know-how is made available at a "price," it may not fully reflect the lost opportunities to the company supplying the know-how; namely, the predesignation of foreign outlets for the processing of oil and export markets, transfer of value-added from the producer countries to the affiliates of oil companies in the ACs, and maximization of retained earnings. In general, oil companies have tended to be relatively less resilient in the transfer of oil technology than

---

cartel. The former originally consisted of the so-called "Seven Sisters"--Exxon, Royal Dutch/Shell group, British Petroleum, Gulf Oil, SOCAL, Texaco, and Mobil. These companies formed consortia in the producer countries and the resultant strong international cartel controlled both the prices and production of oil unilaterally until OPEC was formed in 1960. The four basic objectives of OPEC initially were to stabilize crude oil prices, pursue the right of producer countries to bilateral agreement on prices, seek and obtain equitable prices, and provide a basis for the unification of members' policies through regular consultation. These objectives had since expanded and regular reexamination and renegotiation of existing contracts as well as plans for active indigenous participation in the industry have now become firmly instituted and accepted by the major oil companies and the international oil industry.

manufacturing industries. Oil technology, particularly in oil recovery, had ensured much higher rate of return on the oil companies'
investment than downstream operations, hence the highly proprietary
nature of this technology. The risk factor, however, is also relatively high. These conditions have tended to constrain both the acquisition of foreign technology and local initiatives in the development and
use of local inputs for the processing of oil and gas.

There are two seemingly facile modes of acquiring technical
know-how in the oil industry. These are (1) "learning-by-doing"
among the indigenous manpower through the use of machinery and
equipment, and (2) planned company skill training programs off and
on-the-job in the application of available scientific knowledge and the
development of oil technology. Each of these parallels the two media
identified above. Generally, the flow of technology into mining and
manufacturing, for example, in DCs tends to be in "packaged" form
leaving no room for active local participation in its technological
adaptation and therefore leaves little room for "learning by doing."
Planned company skilled training programs have also been strictly
controlled and the tendency has been to concentrate on the training
that the companies consider conducive to their efficient management
and in their corporate interest. Both modes have only become more
apparent since 1970, and the changes they are undergoing have been
compelled by government policy rather than by the choice of the foreign oil companies.

According to oil mining leases and prospecting licenses issued
since 1969, oil companies are required to employ qualified Nigerians
and train indigenous manpower to replace expatriates. Nonetheless,
the industry's contributions to the development of skilled Nigerian
manpower has been incidental and the transfer of know-how to the oil
and nonoil sectors through the employment and training of Nigerians
has been limited in scope.* Out of the 2,900 Nigerians employed in
the oil industry in 1966, about 500 were in high-level skilled jobs and
this employment category usually includes technicians. Complete information about the entire industry is very difficult to obtain; but, by
relying mostly on the available data from Shell-BP, the largest and
longest established company in the industry, the following pattern is
instructive. The number of Nigerians employed in the industry as a
percentage of the industry's total employment has remained virtually

---

*According to the Nigerianization Decree No. 51 of 1969, oil
companies are required to have Nigerians consist at least 70 percent
of their employees at all operations within seven years of the companies' operations in Nigeria, and rapid increases in this proportion
was also mandated with the ultimate goal of complete indigenization
of the industry.

constant at about 84 percent throughout the 1960s, mostly in the unskilled work force. In 1967, Nigerians accounted for about 47 percent of the 1,318 employed in the high grades positions, supervisory
or managerial and professional work force including skilled technicians. Table 7.14 shows the functional description of employment of
Nigerians and expatriates in 1967, and the dominance of the latter
group in the three high grades. Progress toward changing this distribution has been slow. In 1970, Shell-BP had only one Nigerian
out of 12 in its top managerial grade. Although the distribution was
less biased in favor of expatriates in the professional and supervisory
grades, there were twice as many expatriates as Nigerians in these
categories. When compared to Shell-BP, most of the other oil companies, for example, Gulf, Mobil, Texaco, Tenneco, Agip, and
Safrap have generally worse records in their employment and training of indigenous personnel. The few exceptions where Nigerians
occupy high professional positions are those where prior training
had been gotten independent of the oil companies. Two examples are
in medical and financial training.

TABLE 7.14

Employment in Nigerian Oil in 1967:
Expatriates and Nigerians

Employees	Total
Total employees	3,901
Nigerian	
Management	16
Professional	141
Supervisory	465
Skilled labor	1,043
Other	618
Unskilled labor	969
Total	3,252
Expatriate	
Management	47
Professional	379
Supervisory	185
Skilled technicians	85
Total	649

Source: Federal Ministry of Information, Annual Report of the
Petroleum Division of the Federal Ministry of Mines and Power,
1966-67, Lagos, Nigeria, 1968.

Technical and other forms of training programs by the oil companies have also failed to coincide fully with domestic industrial development programs. The general nature of this assymetry is discussed in chapter 1. Nigerianization as a vehicle for accelerating the time structure of training programs, and for the transfer of decision making and executive responsibility to Nigerians have been recalcitrated by oil companies. This posture has been defended by calling attention to the lengthy process entailed in the training of indigenous personnel. Contributions by oil companies to the development of scientific and technical education consist of periodic grants to educational institutions--universities and technical schools--and company technical and management-supervisory training programs. *
Again, in comparison to all the other oil companies in Nigeria, Shell-BP, the largest employer in the industry, has a relatively comprehensive personnel training program for the development of skills in areas such as plant operation and instrument fitting, and drilling rig machinery operation. Table 7.15 shows the pattern of scholarship awards and training programs for the Nigerian staff of Shell-BP inside and outside Nigeria since 1970. The technical and management training programs are job specific, generally periodic and short in duration. Unlike the slow development of these programs in the 1960s, some growth has become evident in the 1970-72 period, but the training program remains job-specific and peripheral. The actual development and instructive application of technological know-how are performed abroad primarily using foreign manpower. Thus, opportunity for learning-by-doing, and the appropriate acquisition, absorption, and adaptation of oil technology have been forestalled. Furthermore, scholarship awards show a lack of definite commitment for significant expansion. There is also a general lack of information on the rate of attrition or percentage of recipients completing the scientific training and the percentage actually returning for employment in the oil industry.

There is a lack of conclusive evidence that the training programs by the oil companies are intended to hasten the replacement of expatriates by trained Nigerians. The general reluctance to rely on Nigerian personnel in oil and oil-related activities and the obvious disposition toward the training of Nigerians for technical and managerial know-how essential for their major production and executive responsibility, are presumed in the best protective interest of the oil companies. Secrecy and loyalty to the corporate interest are seen as

---

*Only 3 percent of the cumulative expenditure by the oil companies in Nigeria between 1965 and 1970 was for indigenous manpower training.

TABLE 7.15

Training and Scholarship Effort, 1970-72
(for Nigerian staff)

	1970	1971	1972
Training			
Numbers attending:			
Technical training, typically 3-12 months (management staff)	73	106	219
Management/supervisory/other, typically 1 week (management staff)	220	415	812
Artisan/other, typically 12 months (junior staff)	236	191	458
Man-days of training received:			
Management staff--actual man-days	11,753	10,810	19,869
Management staff--as a percentage of total available working days	12.2	8.7	12.8
Junior staff--actual man-days	27,423	35,247	34,770
Junior staff--as a percentage of total available working days	8.2	8.2	8.1
Scholarships			
Numbers attending:			
University, typically 3-4 years	120	124	110
Technical college, typically 2 years	45	38	32
Marine scheme, typically 7 years	3	6	9
Industrial training scheme, typically 3-12 months	--	4	20
Total	168	172	171

Expenditure
Total expenditure on training and
scholarships, 1970-72 .... ₦3,049,200

Other figures
Number of management staff engaged
in training duties at end 1972 52
Total (off-job) training received during
1970-72 equivalent to:
  Management staff .... 170 man-years
  Junior staff .... 390 man-years
Total scholarships awarded, 1956-69 394
Total scholarships awarded, 1970-72 191

Source: Shell-BP Petroleum Development Company, Nigeria Limited, A Report on Training, 1972.

sources of conflict with national loyalty of Nigerians in the international operations of oil companies.

Notwithstanding the amorphous nature of Nigerianization legislation, a reversal of oil companies' proclivity to expatriate employment in the high-grade jobs has become evident since 1970 as a result of the 1969 Nigerianization Decree and its apparent influence on oil companies' employment decisions. But expatriate manpower in the Nigerian oil industry remains far more experienced than the indigenous manpower. According to a Shell-BP study in 1972, the average years of experience for expatriates and Nigerians are 12.5 and 4.5 years respectively.[9] The supply of indigenous scientific and technological manpower has been showing gradual improvement since 1970. This has been greatly facilitated by the establishment of the NNOC in 1971. Compared with past practices, NNOC's equity and managerial participation in the oil companies has made the search for qualified Nigerians and the appropriate training of those currently employed more programmatic. The general posture of NNOC also appears to offer incentive for the inflow of appropriate technical know-how from downstream operations, particularly the processing of oil and gas.

Various aspects of downstream operations, including the processing of oil and gas, are among the major capital investment projects in the Second Development Plan 1970-74. But their implementation was constrained initially by shortfall in revenue and subsequently by technological factors. These projects have recently been carried forward to the Third National Development Plan 1975-80. Some of these projects will be discussed.

One important concern that ought to be of primary consideration in the formulation of national policies on the oil industry should be how to promote the activities of NNOC without its encroachment on the oil companies' "efficiency" and confidence, and thereby strengthen Nigeria's bargaining position in technology acquisition.

There is distinct evidence on the limited access to the acquisition of technology and the exclusion of participation by a broad spectrum of indigenous entrepreneurs in oil and oil-related activities. It is still too early to ascertain empirically how acquisition of equity has had impact on the pace of acquisition of know-how and the development of local technology in the oil industry. One recent development that does not appear to have any apparent relationship to the changes in public policy on the oil industry is the unique circumstances of off-shore west Africa that prompted the development of a special technology. The new technology is a combination for new jack-up off-shore vessel for pipe laying, drilling, and construction. In addition to drilling in medium depth waters, it can retract its legs and be moved into the shallow waters on the west coast of Africa with a light-ship draft of only six feet.[10] For the next five years at least, increases in indigenous participation are likely to be gradual as the shortage of scientific, technological, and entrepreneurial manpower continues to pose serious constraint on local efforts in the growth of the industry. In contraposition to the government's success in output regulations and its influence in raising the revenue from oil, its influence in accelerating the pace at which appropriate technological know-how is acquired and in diversifying the sources of the know-how remain problematic.

On the basis of these past experiences and the current posture by oil companies on the supply of oil technology, the real long-term technological development of the industry and its contribution to Nigeria's overall industrialization can be appropriately achieved through the direct use of the financial benefits from oil. This could be accomplished through the disbursement of government revenue, about 90 percent of which currently comes from oil. Recent developments and plans intended to accomplish these objectives are discussed in the next section on national oil policies.

NATIONAL OIL POLICIES

Major changes in the Nigerian government's oil policy that took effect in the late 1960s included significant legislative actions, but the oil industry did not begin to acquire a posture devoid of colonial guidance until 1970. Among the major changes are the supersession of concession agreements by participatory arrangements; bilateral determination of crude oil prices;* the treatment of royalties and similar payments as part of production cost rather than as part of government taxes as was previously the case; cuts in tax allowances to oil companies; a general increase in the role of the federal government in the control of the industry; and the establishment of the Nigerian National Oil Corporation (NNOC) in April 1971. This is no doubt the most significant change in the history of the industry.

The Nigerian National Oil Corporation (NNOC)

Active indigenous though quasi-ownership participation in the oil industry was formally outlined for the first time in the 1969

---

*There has also been expressed interest by Nigeria and the other OPEC countries to devise means of indexing crude oil prices to take account of the effect of international inflationary trends on their import costs.

Petroleum Decree. But it was not until 1971 that the Nigerian government established the NNOC as a state corporation for indigenous equity participation in the oil industry. Before the establishment of NNOC, oil concessions and drilling rights were based on the Mineral Oil Ordinance 1959, and the earlier guidelines and arrangements set up by the colonial government. These provisions particularly lacked the incentives for the oil companies to want to promote the active linkage of the oil industry with the rest of the economy. Although government revenue and other financial benefits from oil had increased significantly before the inception of NNOC and Nigeria's membership in OPEC, the country had occupied a less advantageous position relative to other oil-producing countries.

The main features of the Mineral Oil Ordinance and the Petroleum Profits Tax Ordinance of 1959, both of which formed the basis of government policy toward the oil industry, show a great deal of generosity toward the oil companies.

These provisions provided major incentives for the growth of oil exploration in the Niger Delta. In general, the provisions also secured considerably less share of oil profits for Nigeria than in Libya, Algeria, and the other member states of OPEC.* Although the inducements in Nigeria had appeared superfluous by the mid-1960s, all the provisions remained in operation until 1967, when the federal military government amended the Petroleum Profit Tax and substituted "posted prices" for "realized prices" as the reference price for tax purposes. All the other salient features of the revisions aimed at increasing the "government take" compared favorably with international and OPEC's standards, and specifically with the Libyan provisions after which the Nigerian amendment was modelled.

The establishment of the NNOC in 1971, empowering it to become involved in any and all phases of the industry--exploration, production, transportation, and marketing of crude oil, refining, marketing of refined products at home and abroad, and the development of petrochemical industries--has caused some changes in the vertically integrated structure of the Nigerian oil industry. In addition to its own oil prospecting rights, the NNOC has been acquiring participation rights in foreign oil companies operating in Nigeria. The pattern of 100 percent foreign ownership in the oil industry was first broken

---

*The Algerian and the Libyan petroleum legislation and agreements differ substantively from Nigeria's. In Algeria and Libya, profits were assessed on the basis of fixed, posted prices; royalties were calculated as current operating expenditure, and capital consumption allowances were considerably lower than the provision of the 1959 Petroleum Profits Tax Ordinance in Nigeria.

in 1971 when the NNOC acquired 35 percent equity in Safrap. By
late 1973 the NNOC had successfully acquired a minimum of 33.3
percent equity participation in 9 out of the 14 foreign oil companies
operating in Nigeria with the prospect that this will rise to 51 percent during a prespecified time frame. See Table 7.16. The NNOC
also seeks 60 percent control in oil marketing operations. Although
the nature of phased participation in Nigeria and the other OPEC
countries is now generally regarded by the world oil industry as
phased nationalization, the Nigerian government has repeatedly expressed disinclination toward complete nationalization. Notwithstanding the significant increase in participation and government control in the industry, the day-to-day decisions on various aspects of
the industry are still made by the oil companies.

The rise in domestic equity and manpower participation since
the establishment of NNOC has entailed the setting of quotas for participation of indigenous managerial/supervisory and technical manpower. But there is still a lack of long-term commitments from the
oil companies to Nigeria's technological development. Although the
"Nigerianization" policy requiring acceleration in the training of Nigerians under new oil mining lease agreements and oil prospecting
licenses has raised the indigenous employment to over 75 percent of
the total employees, the expatriate staff still accounts for a disproportionate share of factor payments to labor.

Furthermore, new oil prospecting licenses issued since 1971
obligated oil companies to provide for educational programs in Nigerian universities in the form of annual contributions or cash grants
for programs related to petroleum technology. Clearly, the crucial
test of the industry's direct and indirect role in Nigeria's industrial
development programs is not only how rapidly it increases employment or output as such, but how rapidly it builds up local entrepreneurship and enterprises. Recognizing that national industrial development depends on linkages among rapidly expanding sectors and
the slow-growth sectors that could benefit from this rapid growth,
public policies are being aimed at promoting the diffusion of technological and managerial know-how from the oil industry to the appropriate sectors of the economy. In addition, one of the important objectives of the Second Development Plan, 1970-74, was the active use
of oil, revenue presumably, to transform the economy into a modern
state, technologically and industrially.

Many of the efforts to control and regulate the activities of the
oil sector in order to optimize its effective contribution to the entire
national economy have been limited to short-term financial benefits.
Available evidence on the singularity of the industry and the stage of
development in Nigeria suggest that both factors have necessitated
the existing technological segregation of the industry and its weak

TABLE 7.16

Petroleum Companies Operating in Nigeria, November 1973

Company	Percent Equity Share	Affiliation
Nigerian Agip Oil (Nigeria)	33.3	subsidiary of the Italian State Company ENI
Phillips Oil Company (Nigeria)	33.3	subsidiary of Phillips Petroleum Company USA
Nigerian National Oil Corporation	33.3	Nigerian (public corporation)
Deminex	30	subsidiary of German oil consortium
NNOC	51	Nigerian (public corporation)
Niger Petroleum Company	19	Nigerian (private) Independent
Japan Petroleum (Nigeria)	49	Japanese consortium
NNOC	51	Nigerian (public corporation)
Niger Oil Resources Co. Ltd.	9	Nigerian (private) Independent
Japan Petroleum (Nigeria)	40	Nigerian (public corporation)
NNOC	51	Nigerian (public corporation)
Tenneco Oil (Nigeria)	37.5	subsidiary of Tenneco Inc., USA
Mobil Producing Nigeria	50	Nigerian (public corporation)
Sun DX Nigeria	12.5	subsidiary of Sun Oil Company, USA
Occidental Petroleum (Nigeria)	49	subsidiary of Occidental Petroleum Co., USA
NNOC	51	Nigerian (public corporation)
Pan Ocean Oil (Nigeria)	n.a.	subsidiary of Pan Ocean USA
Delta Oil	n.a.	Nigerian (private) Independent
Safrap (Nigeria)	65	subsidiary of French Elf Erap group
NNOC	35	Nigerian (public corporation)
Shell (Nigeria)	32.5	subsidiary Royal Dutch Shell, UK, Netherlands
British Petroleum (Nigeria)	32.5	British Petroleum Oil Co., UK
NNOC	35	Nigerian (public corporation)
Henry Stephens and Sons	49	Nigerian (private) Independent
NNOC	51	Nigerian (public corporation)
Texaco Overseas Petroleum (Nigeria)	50	subsidiary of Texaco of USA
Chevron Oil (Nigeria)	50	subsidiary of Standard Oil of California, USA
Gulf Oil (Nigeria)	100	subsidiary of Gulf Oil, USA
Mobil Producing Nigeria	100	subsidiary Mobil Oil Corporation, USA
Phillips (Nigeria)	100	subsidiary Phillips Petroleum Co., USA
Great Basins Petroleum Dev. Co.	100	Nigerian (private) Independent

n.a. = not available.

Note: Companies with Nigeria after their names are incorporated in Nigeria, in most cases, under the name of the parent company.

168

linkages with the rest of the economy. But the increasing equity and manpower participation by NNOC offer definite opportunities for the promotion of backward and forward linkages. The NNOC's policies and initiatives to promote long-term technological benefits from oil and facilitate industrial--manufacturing and agricultural--development are further explored in the following discussion of its role in exploration and production, downstream integration and the development of oil-based industries, and manpower training and technological capabilities, including management/organizational and marketing know-how. All three phases have definite adaptive and linkage capabilities with other sectors of the economy.

Exploration and Production

Since it was established in 1971, NNOC's participation in upstream activities has inevitably been through cooperation with foreign investors and potential suppliers of suitable technological know-how.

The NNOC search for technology essential for exploration and production began in 1971. Among the sources from which appropriate technology could be acquired presumably under nonproprietary arrangements, and from which probable cooperation has been indicated, are the governments of Algeria, Rumania, and the Soviet Union. Cooperation from private foreign oil companies in providing the appropriate technological know-how under nonproprietary arrangements has been rare. Although the 1973 NNOC/Safrap venture, a cooperation for exploration, production, and refining of oil, involved a $4 million U.S. Eximbank and Chase Manhattan loan for the purchase of U.S. oil equipment, it fell short of providing appropriate technology. The venture was limited to the purchase of oil equipment.

The NNOC continues to survey the market for technological know-how required in exploration and production operations while also conducting manpower recruitment for its exploration and production departments, with heavy concentration on geologists, geophysicists, petroleum engineers, reservoir engineers, and production engineers. One area in upstream operations in which the NNOC has made definite inroad is in seismic survey.

The apparent irreversibility of a substantial part of the rise in oil prices in 1973 and 1974 should be expected to result in a shift by foreign oil companies from complete vertical, upstream-downstream, integration toward downstream integration, as the larger proportion of the profitability of upstream operations that had accrued to oil companies earlier, shifts to the governments of the producer countries. Furthermore, oil companies' participation in upstream activities is being limited directly. In the Nigerian case, the government decided in 1972 that all future concessions will be vested in the NNOC. However, the shift toward downstream integration by oil companies is likely to be gradual as the higher crude oil "prices" improve the outlook for technological developments in more advanced oil-recovery techniques. To fill the void that is likely to develop in upstream operations would require a systematic buildup of operational capability by the NNOC, and this is primarily dependent on access to appropriate technological know-how that potentially exists through bilateral arrangements with countries like Algeria and the Soviet Union.

Downstream Integration and Oil-based Industries

Since its establishment in 1971, the NNOC has been able to establish the desirability of expanded refinery operations in Nigeria on the basis of increasing internal and external product demand to meet the energy-intensity of worldwide industrial development. Although the price of crude oil quadrupled in 1973/74 and the world demand for oil products remains precarious, increased demand is imminent as the market price of crude oil stabilizes and the unusual Middle East political ferment that has been responsible for oil embargoes subsides. In Nigeria, the domestic consumption of refined oil products is growing at an annual rate of 9 percent, and a similar trend is expected in most of the other west African countries at least through 1980.[11] Nigeria's domestic market itself accounts for about 2.5 million tons annually. This market is shared by a few foreign oil companies and their affiliates, and their duplication of overhead has meant that relatively high transfer prices for oil products prevail in the local market. * This problem has been compounded periodically by the limited capacity of the one refinery in Nigeria. The refinery built at Alese Eleme near Port Harcourt is 60 percent government interest and 40 percent Shell-BP. It began production in 1965 and has a capacity of 1.5 million tons annually. Although it has expanded its capacity, it is unable to meet the growing demand for refined oil products. Two new refineries are planned for Warri and Kaduna, but they are not expected to go into production until 1976 and 1980 respectively. When completed, both are planned to have capacity for export. Apart from the limited access to the export market

---

*The individual market share of the seven oil-marketing companies are as follows: British Petroleum (BP), 32 percent; Shell, 23 percent; Mobil Oil, 18 percent; ELF, Esso, and Texaco, 8 percent each; and Agip, 3 percent.

for refined oil products, there is also the primal problem of acquiring appropriate refinery technology. In 1971 the NNOC commissioned
a study of the alternatives of refinery operations and the general supply of refinery technology. As a result of this study, France extended
a $20 million credit to Nigeria for refinery and petrochemical proj-
ects, but the possibility of acquiring appropriate oil refinery technology was not considered. * In its investigation and search for alternative sources of technology, the NNOC could benefit from the lessons
learned by other DCs in the last decade. These were lessons from
the development of indigenous refinery operations, attempted importation of refinery technology, particularly experiences about the central
role of "terms of transfer" in effectively acquiring the technology,
and the problem of scale economies.† Usually, contracts with foreign firms for the establishment of refineries in DCs cover a broad
spectrum from plant location to pricing of petroleum products for the
domestic market. Although incremental cost of output expansion is
relatively low, this is commonly discouraged. Where export capacity
is part of a contract, the marketing and pricing of the residual output
in the export market are closely controlled by the oil companies. Unlike numerous other DCs seeking to establish oil refineries, Nigeria,
through the NNOC, possesses its crude oil supplies and has had about
a decade of experience in oil refining locally. Theoretically, it occu-
pies a strong enough bargaining position to avoid "packaged" technology agreements typical of international technology flows, but the proprietary nature of the suitable technology puts its suppliers, particularly the oil companies, in a much more advantageous position.

The Second National Development Plan 1970-74 had committed
the government to hydrocarbon projects including refineries, petrochemical, and natural gas projects. This has necessitated the direct
involvement of NNOC in downstream operations in addition to its participation in upstream operations. As is generally characteristic of
industrial development in DCs, these projects require joint-venture
and are therefore faced with potential restrictions about capacity,
operational control, and export markets. These restrictions have
so far forestalled the development of Nigeria's planned petrochemical
complex, as major oil and chemical companies in the West, including
Shell-BP, surreptitiously resisted its establishment, with the proposed

---

*One example of appropriate know-how in downstream opera-
tions is refinery technology for different types of oil products from
the same crude oil.

†Modern refinery techniques, for example, catalytic cracking,
involves problems of indivisibilities and thus requires a high rate of
throughput for efficient production. Location problems aside, Nigeria
offers a viable market.

export capacity. The project has been looked upon unfavorably by
major oil and chemical companies on the grounds that the world market for petrochemicals is well-supplied by existing plants. This
position is understandable since these plants are, in many cases,
owned by vertically integrated oil companies and their affiliates.
Invariably, this posture has impeded integrated oil companies and
their affiliates. Invariably, this posture has impeded the "free"
flow of appropriate petrochemical technology. The situation is much
the same for projects based on natural gas, except in this case, the
development of large refrigerated liquid natural gas tankers has
opened new export opportunities to producer countries and potential
foreign partners are anxious to participate with the Nigerian government in the production and export of liquefied natural gas and liquefied petroleum gas to the U.S. market, where demand is high and
rising. Nigeria has vast reserves of natural gas in addition to those
associated with oil drilling. Locally, natural gas is a potential
source of low-cost energy; but gas liquefication is still a compara-
tively new technology. Natural gas represents a vital source of energy for the development and growth of natural gas intensive industries such as petrochemicals and nitrogenous fertilizers including
ammonia and urea. These assessments are supported by the generally favorable results from studies so far conducted to assess the
viability of Nigeria's natural gas for the development of specific
natural gas intensive industries. One such study was conducted in
1964 by Canadian Industrial Gas Company Ltd., and recommended,
among other things, the establishment of six such industries--liquid
product recovery, polyethylene, fertilizer, carbon black, salt production, and caustic chlorine.

     Despite its viability and the most economically potential feed-
stock for electricity, in the early 1960s, the Nigerian government
became heavily indebted to the World Bank for loans to build the
hydroelectric Kainji Dam project while natural gas was being flared
negligently during oil drilling operations. Clearly, the Dam project
lacked rational economic consideration, and the apparent incomplete-
ness of the World Bank's study as well as the dominance of sectional
political influence in Nigeria both obviated the full exploitation of
natural gas in spite of its potential as an alternative and cost-saving
source of energy.

     The flaring of about 2 billion cubic feet of natural gas daily in
Nigeria during the late 1960s and early part of the 1970s represents
approximately 17 percent waste of the country's total daily energy
production. Two projects on the production of liquid natural gas
(LNG) have been included in the Third National Plan. These are two
separate government joint ventures with Shell-BP and Agip-Phillips.
Both projects and the planned fertilizer complex are indicative of the

final recognition of the need to harness the appropriate technology and put natural gas to economic use.

Opportunities for further development and expansion of oil-based industries have also become more and more apparent, particularly for agro-allied activities as a result of new developments in the exploitation of natural gas. For example, fertilizers are largely based on petroleum feedstocks and represent the single most important factor in increasing agricultural production. This offers the opportunity for a vital linkage with the agricultural sector. World consumption of commercial fertilizer has been rising rapidly since the early 1960s and it soared from 49 million tons in 1965 to over 76 million tons in 1969. Further spectacular increases are expected throughout the 1970s and 1980s. For Nigeria, this offers opportunity in an important export market. Similar opportunities also exist in domestic and export markets for oil-related construction materials and other animal feed, but the existence of economies of scale is essential if the industry is to take advantage of the export opportunities. Typically, fixed costs in petrochemical production decline rapidly as plant size is increased. It is therefore advantageous for production to be capable of expansion in order to cater to the sizable and growing domestic market and for the likely expansion in export market. The requirement for large-scale capacity expansion in petrochemical production inherently produces export capacity, but changes in the consumption of the industry's products are normally gradual. Apart from periodic capacity shortages in the ACs in petrochemical production, export possibilities for DCs are in general substantial but barriers to entry remain firmly in effect. [12]

Manpower Training and Long-term
Technological Development

The effective acquisition of suitable foreign technology and the development of indigenous know-how depend neither on chance nor isolated incidences. They depend on the constant and systematic flow of scientific and technological information. To ensure such a flow would require adequate financial resources as well as the establishment of trust in the supply and acquisition of technology, stable but open long-term relationships between the suppliers of technological know-how and the recipients, and between national research institutes and the ultimate user of the output of research, industry.

In recognition of these preconditions, the NNOC has chosen a programmatic approach in correcting existing deficiencies in indigenous manpower training and specifically in improving local technological capabilities in upstream and downstream operations. Two

cases in point are, the high priority the NNOC has accorded research
and development locally and the establishment of the Training Institute for Petroleum and Petrochemical Industries by the government.
The Nigerian government has also established in 1975 a teaching center in Warri through technological collaboration with the Soviet government to provide training for Nigerians in oil technology.

The highly developed organizational and marketing know-how
in the oil industry also offers potentially beneficial spillovers through
cooperation between foreign oil companies and NNOC in the training
and development of Nigeria's entrepreneurial manpower. This requires, for example, identifying applicable and facile accounting practices and management techniques in the industry and adapting these to
the conditions and needs of Nigerian entrepreneurs. In sum, the success of the search for alternative sources of oil technology is very
much dependent on appropriate manpower training for the establishment of the necessary technological capabilities that could facilitate
the search and the effective use of the acquired technological know-
how.

CONCLUSION

The vertically integrated structure of foreign firms in the Nigerian oil industry has fostered conditions inhibiting the "free" and
effective transfer of technology. Although no direct causation has
been established between these conditions and the acquisition of oil
technology in Nigeria, the ability of the companies to maintain exclusive rights in the use of their technological know-how appears to
have been facilitated by the integrated structure of their operations.
Recent changes in national policies toward the oil industry, especially the participatory relationship between the government and the oil
companies, offer some basis for another form of integration that has
potential benefits; namely, the proper integration of the industry with
the rest of the economy. Of course, there is no automatic link between government participation or majority ownership and effective
technology transfer.

There is no conclusive evidence that the acquisition of equity
in and of itself has had any apparent impact one way or the other
either on the pace of acquisition of technological know-how or the development of local technology in the oil industry. The final measure
of positive changes resulting from government participation in the
industry depends on NNOC's effectiveness in the acquisition of technology, the transfer of technology from the industry to the other sectors of the economy, and on the extent to which Nigerian entrepreneurs
become involved in the future development of oil and oil-related industries.

In comparison to Nigeria's weak position in its bargaining with the oil companies in the 1950s and 1960s, the NNOC since its establishment in 1971 has occupied a strong bargaining position in its search for oil technology. This position is likely to get stronger as a result of its overall participatory posture, and a continuing relatively high world demand for oil and oil products.

Nigeria's capacity to acquire and absorb suitable technology and develop indigenous ones should be enhanced by the NNOC's policy to promote partnership and licensing agreements with willing suppliers of technology, and diversify sources of technology; concentrate on building up local manpower and R and D capability; increasing local technical competence, a factor that may have been underestimated; and continually court the increasing number of foreign governments that are potential suppliers of oil technology. Although the up-to-date know-how and technological innovations are in the United States and Western Europe, appropriate oil technology capable of meeting the needs of producer nations like Nigeria and Venezuela are also available, for example, in India, China, and the Soviet Union.

In the NNOC's build-up of indigenous technological manpower, a shift from the planned rapid expansion in high-level manpower development in 1972 to short- and medium-term development of middle-level skilled technicians is essential for the effective acquisition of foreign technology, because the latter group tends to adapt more readily than the former in relating imported technology to the prevailing national conditions and needs.

The future of the oil industry and its contributions to industrial development, agricultural and manufacturing, will depend on the practicality of national science and technology policies and strategies. The effectiveness of the use of income from oil, which accounts for about 90 percent of government revenue in the last quarter of 1974, in the acquisition of equity and manpower participation at all skill levels for the development of indigenous manpower and know-how also depend on these overall policies and strategies. The sober fact that oil reserves are finite sounds an important imperative note for the rapid expansion in financial benefits, but the optimum plan is the one that includes financial as well as long-term technological benefits. Given that effective technology transfer and acquisition has not been a corollary of DFI in the Nigerian oil industry, public policy must ensure that financial resources from oil are directed to programs and projects capable of promoting long-term technological benefits for national industrial and agricultural development. Industrialization (manufacturing and agriculture) is, of necessity, an alternate source of income and employment after the depletion of oil reserves.

NOTES

1. See L. H. Schatzl, Petroleum in Nigeria (Oxford: Oxford University Press, 1969), and S. R. Pearson, "Nigerian Petroleum: Implications for Medium-Term Planning," in Growth and Development of the Nigerian Economy, ed. C. E. Eicher and C. Liedholm (Michigan: Michigan State University Press, 1970), pp. 352-75.

2. Scott R. Pearson, Petroleum and the Nigerian Economy (Stanford: Stanford University Press, 1970).

3. Schatzl, op. cit., p. 94.

4. Standard Bank Ltd., London, Standard Bank Review (September 1974), p. 17.

5. See Federal Republic of Nigeria, Second National Development Plan 1970-74, and Third National Development Plan 1975-80, Federal Ministry of Information, Lagos, Nigeria, 1970 and 1974.

6. Michael Tanzer, The Political Economy of International Oil and the Underdeveloped Countries (Boston: Beacon Press, 1969), p. 128.

7. In a study of the Nigerian manufacturing industry, this practice was found to be the rule rather than the exception. See D. B. Thomas, Capital Accumulation and Technology Transfer (New York: Praeger, 1975), chapter 3.

8. Pearson, Petroleum and the Nigerian Economy, op. cit., p. 57.

9. Shell-BP Bulletin 7, no. 12 (December 1972), Lagos, Nigeria.

10. See Oil and Gas Journal, March 3, 1974, p. 104.

11. Petroleum Intelligence Weekly, May 7, 1973.

12. See R. B. Stobaugh, The International Transfer of Technology in the Establishment of the Petrochemical Industry in Developing Countries, UNITAR Research Report No. 12 (1971), Unitar, N.Y.

8

GENERAL CONCLUSIONS
AND RECOMMENDATIONS

In the preceding chapters, major questions have been raised about the contributions of direct private foreign investment to the economic development of African countries and DCs in general. The direct and indirect "technological components" of these contributions have been of particular interest in the various analyses of the role of DFI. The broad historical perspectives developed in chapters 1 and 2 served as the foundation for these and other analyses of the major issues covered in the subsequent chapters. These major issues include DFI-related conceptual problems, and the examination of available empirical evidence on the importation of technology to a number of African countries through DFI. In this chapter, the previous seven chapters are summarized briefly and their conclusions identified. Some broad conclusions are drawn from the entire volume, and a few policy recommendations are made.

Although the empirical evidence examined in the foregoing chapters is incomplete, there is a definite suggestion that the conceptual problems associated with the flow of DFI and the importation of technology by one AC from another are likely to be different from those involving flows from an AC to a DC. The significance of this difference becomes apparent in those instances where the subjects of empirical research involve the differences between ACs and DCs in scientific and technological development, and other essential but often neglected questions about how to identify a successful "transfer" of technology from ACs to DCs, and what the criteria are for the measurement of the success.

Satisfactory answers to these questions are fundamental for the proper description of a conceptual framework. The importance of these questions initially prompted the distinction in chapter 1 between technology "transfer" and technology acquisition. For illustrative

parallel, the assertion is that the former is supply push in character, and the latter is demand pull in nature. These distinctions are
considered instructive on why technology "transfer" as a concept
may not be applicable in the case of DCs because in its common
usage, the success of the transfer process is usually presumed.
Therefore, in the case of flows from ACs to DCs, an important condition for successful technology "transfers" is the active involvement
of both the suppliers and recipients of the technology in all phases of
the transfer process. However, such involvement by recipients requires appropriate and relatively developed technological infrastructure. This is perhaps the major determinant of most DCs' absorptive capacity for the beneficial use of imported technology. The extent to which relatively undeveloped technological infrastructure may
have constrained the absorption of imported technology in African
countries may be ascertained qualitatively at least by comparing what
the "technological components" of specific DFIs were capable of contributing with what was actually contributed in these countries. Clearly, the infrastructural requirements are elemental to the necessary
adaptation and acquisition of the imported technology, and more importantly, for the extension of the inflow beyond the mere supply of
technological information.

In chapters 1, 3, and 5 some of these requirements were examined. The conclusion is that "abstract technology" as a part of an integrated flow of DFI has major limitations in promoting long-term
technological development in African countries. The major components of these constraints are a relatively undeveloped technological infrastructure, the supply-push characteristics of the technology
inflows and its misdirection from aiding fully in the development of
appropriate scientific and technological infrastructure, and in adaptive and innovative industrial programs, instead of duplicative ones.
Some of the requirements for the minimization of the constraints are
discussed specifically in chapter 3.

Notwithstanding the stage of technological infrastructure development, it is essential that demand-pull forces influence the supply
of technology in direct response to the expressed needs of the importing country. Demand-pull forces are being evidenced in a few African countries since the early 1970s by the new assertiveness on the
conditions for future participation of DFI and, hence, its potential
technological contributions to economic development in these countries. The importance of active involvement by suppliers and recipients in technology flows is also underscored in Theodore Schlie's
discussion of "vertical transfer" in chapter 6. The case study in
that chapter will be examined briefly here.

Apart from the infrastructural requirement as a condition for
effective technology acquisition, the fundamental economic forces

that bring the suppliers and the recipients together, particularly through DFI, are likely to determine the degree of the effectiveness of the recipients' involvement in the flow process.

Historically, the inflow of DFI into Africa has been in response to three broad factors: (1) the availability of essential raw materials; (2) liberal foreign investment policies throughout the continent primarily by the design of the colonial government to subserve and promote its commerce (in most cases these policies have continued long after the attainment of political independence in many of these countries); and (3) since the 1960s, import substitution policy and the need to overcome some of its presumed adverse effects on the commodity exports of ACs. However, the volume of the DFI inflow has been small when compared to South America or Southeast Asia, and the reasons for these differences have been geopolitical in nature instead of predominantly economic. The development experience in African countries suggests that the inflow of DFI-related technological innovations in the last two decades has not been motivated primarily by the real development needs of these countries. Rather, it has been motivated largely by the MNCs' search for new resources and new markets. Consequently, the inflow has had a predominantly supply-push orientation. The inflow of DFI itself has been selective, with the major concentration initially in trading companies and the "extractive" industries.

The accompanying inflow of foreign technology has been, first through the adoption of new imported commodities, and subsequently through the gradual development of local operations and the adoption of new processes. The experience in the continent seems to support the analytic framework in chapter 1, which described the sequence in which the three primary modes for technology flows are likely to be operative in DCs; namely, export commodities as the initial mode, followed by DFI, and licensing arrangements. The primary consumer market orientation in the flow of export commodities to African countries considerably limits the effectiveness of this channel for the "transfer" and acquisition of foreign technology. Given the appropriate conditions, the flows of technological know-how through DFI and licensing agreement have greater potential mutual benefits for suppliers and recipients than exports. In the case of licensing, its desirability for DCs depends on the bargaining power of the parties to the licensing agreement. Ordinarily, a foreign firm would resort to licensing in a market that could not be reached by exports or through DFI. Apart from the relatively weak bargaining position of many African countries, the use of the licensing mode for technology acquisition has been significantly constrained by the relatively undeveloped scientific and technological infrastructure in these countries. Consequently, DFI, principally by MNCs, has been and will continue to be the dominant mode, at least for the immediate future.

The historical perspective provided by Mira Wilkins in chapter 2 covers the various commercial activities in which DFI has participated in African countries over the years. Namely, retail trade, banking and insurance, mining including oil, agriculture ventures, and multinational manufacturing operations. In virtually all these activities, the initial contacts for the participation of DFI were South and North Africa. In these experiences, the primary contributions of DFI to African countries include social organization, new forms of administration, and technological change from subsistence farming to cash crops, and technical skills from assembly operations. Furthermore, portfolio investment in mining companies contributed to the introduction of foreign management and control, and therefore served as conduits for the importation of technology into African countries. An important research question that arises from this experience is, what were the roles and contributions of financial institutions in the "transfer" and possibly the diffusion of technological innovations? Typically, expatriate investments in the principal sectors--mining, agriculture, manufacturing, and financial institutions-- were related to the imperial-colonial power in the individual countries.

In those instances in which mining and agriculture ventures have provided the basis for skill training, Wilkins concluded that this may have also improved the techniques in use, but there is no evidence that the technology was absorbed and routinized locally. In addition to skill training, these ventures also involved investment in infrastructure and infrastructure-related technology. An example is the construction of railroads by Lever Brothers in the Congo and roads by Firestone in Liberia. Although Africans learned certain skills and work habits, infrastructure-related technology, for example, in communications and construction, was not transferred, but remained in expatriate hands in the same manner that mining and agricultural technology was not acquired and absorbed. In general, all these activities failed to generate effective imitation. With some reservation, the historical trace of events led Wilkins to the conclusion that some technology "transfer" did take place in a number of instances, but that the absorption of the technologies had been unduly slow. Further but similar historical work is necessary for the compilation of evidence that could provide conclusive results.

Africa's natural resources, real and potential, are serving more and more as major attractions for DFI. When the magnitude of these resources is compared individually and collectively with its population's relative material poverty, more questions are raised than resolved, particularly about the "costs" to the continent relative to the "benefits" from the exploitation of these resources through DFI.

A sample of its natural resources based on the world's known reserves puts the continent in a significantly advantageous position in terms of its share of the world's land surface and population, which account for about 25 percent and 10 percent respectively. For example, its share of vital resources out of the world's reserves includes 42 percent of cobalt, 34 percent of bauxite, 17 percent of copper, and one of the world's major iron reserves.

The optimization of "benefits" from the exploitation of these resources by foreign investment has justifiably become a national concern in an increasing number of African countries. Chapter 3 focused on this concern and the public policy changes aimed at its source. One important outcome of this concern has been the increasing search and the development of measures considered necessary to take the fullest possible advantage of DFI by the use of systematic project appraisal, periodic duty revisions, and withholding taxes to overcome loss of revenue as a result of transfer pricing. More importantly, a measure gaining in importance in an increasing number of these countries is the institution of investment codes to control the inflow of DFI and its functional destinations. These codes represent policy reactions to the concern of DCs in general on the pattern of ownership, control, and the perceived cost of DFI in terms of inequitable compensation for exploited material resources and profit repatriation by foreign investors. African countries with a comprehensive science and technology development policy or foreign investment policy--or both--are still few, and the use of investment codes for the screening of DFI to ascertain their economic, political, and social desirability is limited to an even smaller group of countries. These are countries with the type of resource endowments that ensure a relatively strong bargaining position in their dealings with foreign investors and suppliers of technology, primarily MNCs. But, the capacity of African countries in general to implement national policy on S and T, and economic development, is very minimal, and their degree of effectiveness varies considerably. Although the new assertiveness by these countries' public policy toward DFI has begun to take effect, as demonstrated by increasing local equity participation, their relatively undeveloped indigenous entrepreneurships have precluded significant local private participation. Consequently, the number of public enterprises and their equity participation in foreign companies have increased considerably. The implications of existing and continuing changes in investment codes and commercial policies for the future growth of DFI and the inflow of foreign technological innovations have yet to be understood. However, partial evidence from a number of countries, for example Nigeria, Zaire, Sierra Leone, Zambia, and Ghana showed no extraordinary decline in the inflow of DFI or any other conclusive adverse effect as a direct result of such policy changes. One explanation for this is that increases in local equity participation have served political and psychological functions rather than provide an effective means of gaining real local control of key industries dependent on local raw materials. This seemed to have been the case, for example, of oil in Nigeria; diamond, iron, and steel in Sierra Leone; and copper in Zambia. This experience is not limited to Africa; it has apparently occurred in Jamaica with respect to bauxite and the aluminum industry.

Further analysis of the supply of technological innovations by foreign companies to DCs was developed in chapter 4, with the primary focus on choice of products and choice of techniques by parent companies and their subsidiaries in the DCs. This provided a direction different from the role of international agencies and domestic policy analysis in chapter 3. The common problem of incompatibility between choice of imported techniques and domestic input structures was explained by the tendency for the choice of product to precede the choice of techniques. Available evidence suggests that this pattern has been motivated by preferences for sophisticated, latest vintage, and typically capital-using technology in most of the stages of plant production since these provide opportunity for the spread of relatively high fixed cost associated with capital-intensive techniques. There are a few exceptions to this proclivity, primarily on the part of MNCs, to use up-to-date technology.

One example discussed in chapter 4 is Howard Pack's study of the Kenyan manufacturing industries, which included paint production, cotton textile, and food processing. One of his findings is that some foreign companies do search for old equipment that is obsolete by production standards in the ACs, but highly serviceable in DCs. He also found evidence that these companies undertake the reconditioning and, presumably, the adaptation of the old machinery to take advantage of their labor intensity and low cost of production.

However, it is important to note again that the incidence of this practice in African countries is still the exception rather than the rule and tends to be industry-specific. To encourage more of this practice, we need a good understanding of what is involved in the choice process, techniques as well as products. Unfortunately, in consideration of choice processes, most discussions of foreign companies' role in the flow of technological innovations to DCs have generally been more notional then empirical, and a great deal of research still needs to be done. In this respect, policy makers urgently need to have a clear understanding of decision rules on choice of products and choice of techniques in order to design appropriate and viable public policy for the reorientation of foreign companies, especially MNCs, toward the major development needs of host countries.

The proprietary nature of commercial technology and the tendency of suppliers to effect their "transfers" in composite forms pose other problems of major proportions. In chapter 5, Walter Chudson

examined case studies of these problems in Kenya and Tanzania. He
also studied public policy responses to them. The arbitrariness associated with composite supply by ACs to DCs, largely through MNCs,
precludes explicit pricing of technology and facilitates the inclusion of
implicit cost and economic rent for exclusive ownership of the technology. The recognition of the necessity for the appraisal of direct
foreign investment on a project basis in Kenya has revealed some of
these practices as well as cases of foreign exchange overvaluation
that resulted in a major divergence between the opportunity cost of
capital and their market prices.

Chudson also found cases involving the importation of inputs at
low cost through the overvaluation of foreign exchange rates and the
importation of some inputs on a duty-free basis. These practices
were in effect simultaneously with tariff protection and import restrictions. Consequently, profits in manufacturing, the primary
sector affected, have been grossly distorted. In the case of Tanzania,
Chudson noted the recognized need for a change in the evaluation of
contributions by DFI to the national economy from the "doctrinaire"
approach, which tends to limit the consideration of the contributions
to foreign exchange, domestic value-added, and employment, to the
use of "shadow prices" for project evaluation and selection, and possibly for the choice of technology. Using the cashew nut case to illustrate the complexity of problems associated with choice of techniques, Chudson underscored the general importance of export markets to DCs. The stimulus of a large-scale export market for Tanzania's cashew nuts helped in mobilizing resources for the choice of
appropriate technology. Experiments with joint-ventures and management contracts as a means of reducing DFI and its control of the
local economy in Kenya and Tanzania while retaining the associated
foreign technology, have yielded different preliminary results.

In the Tanzanian case, nationalization of foreign enterprises
following the Arusha Declaration in 1967 left some room for joint-
venture as well as management contracts with the objective of obtaining technological and managerial know-how at the least possible
cost. Public policy in Kenya has been more flexible than in Tanzania.
However, efforts in both countries in the monitoring of DFI inflow
are currently being directed toward systematic project appraisal.
Based on Chudson's study, both countries appear to be taking a more
and more flexible posture in relation to foreign business participation, but continue to scrutinize contractual arrangements to ensure
that foreign participation consistently reflects national interests.

The contribution by Theodore Schlie in chapter 6 covered a
topic entirely different from those covered in the other chapters.
It is a case study of some of the coordination of research efforts
among research institutes in the "vertical transfer" of technology

among the countries in the East African community. The study focused primarily on the East African Agricultural and Forestry Research Organization (EAAFRO), a regional scientific research institute based in Kenya, and its relationship to national agricultural and forestry research systems in the member countries of EAC, namely Kenya, Uganda, and Tanzania. Schlie attempted to explain the problems raised in the analytical framework in chapter 3 concerning regional scientific research institutes as vehicles for the maximization of investment in research through economies of scale. In all the various manifestations of regional cooperation, including scientific and technological cooperation, the development of appropriate methodology for benefit-sharing has always been a major obstacle. Furthermore, the precise role of the regional institution in relation to the national systems supporting it has been a major source of confusion. In order to measure scientific benefits from EAAFRO and the benefits being distributed among the member countries of EAC utilizing its services, Schlie made some definitional distinctions between research results and scientific services as a means of identifying two different forms of benefits from EAAFRO's scientific research, and used field interviews of both the "senders" and the "receivers" to determine the perceived benefits by the latter.

The "locale specific" nature of agricultural and forestry research, which involves ecological factors like altitudes, rainfall patterns, and soil conditions, were found to constitute major limitations on the "transfer," the applicability and maximization of benefits from the results of research. The location of EAAFRO in Kenya explains, in large measure, the relatively more effective "transfer" between the organization and national research institutes in Kenya, in comparison to Uganda and Tanzania. The benefits included experimental collaboration involving overlap and complementarity of projects, and scientific and advisory services. The overlap in EAAFRO's coexistence as a multinational research body with national research institutes in the three countries served by its operations is one of the other factors cited by Schlie in the limited transfer of technology between EAAFRO and these national research institutes. A major lesson from this study is that the questions about the roles of regional research institutes and the benefit sharing of their research results are clearly interdependent.

Attempts in chapter 7 to test empirically the proposition that the lack of technological "benefits" from upstream and downstream operations in the Nigerian oil industry significantly constrained the total impact of the financial benefits from oil on the Nigerian economy proved futile because of the inaccessibility of essential statistical data.

From the partial empirical evidence examined, several factors were delineated as highly likely constraints and capable of minimizing

technological benefits from the oil industry. Among these factors
were: the sheer dominance of DFI and invariably the complete foreign control of the industry before 1971, and the primary interest of
the oil companies in crude oil extraction and its export to Western
Europe and North America; the vertically integrated structure of the
foreign firms in the industry and the preclusion of the opportunity for
the development of oil-related service industries through local participation; the conduct of research and development exclusively outside Nigeria to protect the company's technological secrecy--a practice that is by no means unique to the oil industry; major reliance on
expatriate personnel usually from the country represented by the
parent company, as a means of protecting the company's commercial
secrecy; shortages of indigenous scientific and technological manpower; the Nigerian government's equivocal public policy on the oil
industry and the development of scientific and technological manpower
between 1960 and 1971 in particular.

     Although the oil companies have provided periodic training pro-
grams for indigenous personnel, these programs lack definite long-
term commitments for substantial expansion in indigenous manpower
participation. Major policy changes affecting the oil industry began
to take place in 1969. These included the Nigerianization Decree
No. 51 and the establishment of the Nigerian National Oil Corporation. These policy changes and the new assertiveness of the Nigerian
government have begun to gradually reverse the past trends in the employment of indigenous personnel and the programmatic training of
local managerial, scientific, and technological manpower. A major
factor in this reversal is the NNOC's direct and active participation
in the oil industry, including oil prospecting, technical training programs, and participation rights in foreign oil companies. The future
of the industry and its contribution to national economic development
clearly depend on national science and technology policies, and strategies developed to integrate the industry with the rest of the economy.

     In this respect, a strategically developed diversification in the
sources for the supply of technological know-how, and the commitment of major resources to local manpower build-up and the development of appropriate R and D capability are fundamental to the effective acquisition of foreign technology. Recent diversification into
transportation, marketing, and the development of petrochemical industries, in addition to exploration and production, is a major initial
step toward the integration of the industry with the national economy.
The next logical phase in these structural changes is the establishment of backward linkages to local industries for the full exploitation
of other raw materials, and forward linkages with a broader class of
final goods-producing industries.

In addition to the foregoing chapter summaries and conclusions, the following general conclusions and broad recommendations are offered. The fundamental asymmetry between the goal of short-run maximization of economic opportunities in African countries by most foreign companies on the one hand, and the medium- as well as presumed long-term development objectives of African countries on the other, has precluded the full exploitation of all the economic development capabilities of DFI including technology. In light of this experience, there have been mounting pressures for increases in labor absorption, local control, and even for some form of mutually beneficial synergy between MNCs and host countries, but these issues are not likely to be resolved solely through the current trend in joint ventures between MNCs and public enterprises. Major collaborative efforts are also necessary in broadening the stages of production in which labor-intensive techniques are economically profitable and socially useful in terms of generating high employment.* Priority areas in such a focus should, of necessity, include agriculture and small-scale industries.

Given the severe budgetary and local skilled manpower constraints faced by African countries, a viable response to the resolution of the issues raised here is regional economic cooperation. Major research efforts, however, are required to identify the appropriate roles for regional economic groupings and develop alternative methodologies for the sharing of benefits among participants.

These problems have been of primary concern in the East African Community and will require continuous consideration in the functioning of the recently established Economic Community of West African States (ECOWAS). The positive efforts that led to the establishment of the latter clearly add a new dimension to regional cooperation in Africa. Among the obvious anticipated developments within ECOWAS in the future are a free trade zone, a large regional market, and growth in DFI by MNCs.

Governments in ACs have the major responsibility of making constructive responses to DCs in terms of the global operations of MNCs by instituting policy measures that will encourage reinvestment of profits and promote liberal policy on the effective and complete "transfer" of technological know-how that, based on the explicit national needs of the host DCs, are capable of facile internalization and routinization. On the part of DCs, this requires selectivity

---

*A note of caution. It is erroneous to consider labor-intensive technology as coterminous with appropriate technology since not all labor-intensive technologies are appropriate in terms of their competitiveness or profitability and social utility.

and avoidance of unnecessary replication of the structure from where the technology originated. Furthermore, it is essential to minimize serious value conflicts in the inflow of technology and control the influence of the political theory of development in the transferor country.

Governments in African countries must insist on the reorientation of MNCs so due consideration is given to host countries' national development objectives. In all formulation of public policies on DFI, a framework for its integration with domestic science and technology policy is vital to its success and to the effective acquisition of imported technology. Foreign technologies are being effectively acquired when payments for routine services, expatriate "experts" and royalties on the technology are considerably minimized, and the technical adaptive and innovative capabilities of the indigenous workforce can sustain the production process undisturbed. Given a relatively well developed technological infrastructure, the technology supply made capable of meeting these requirements is licensing.

It is evident from the foregoing considerations that the success of the policy measures discussed will depend significantly on the ability of all DCs to implement them based on crystallized national development objectives.

Advisory Committee on the Application of Science and Technology to
    Development. "Third Report to the Economic and Social Coun-
    cil." United Nations. E/4178, May 1966.

Asiodu, P. C. "The Future of the Petroleum Industry in Nigeria,"
    Seminar paper. Lagos: Ministry of Mines and Power, March
    1971.

Canadian Industrial Gas Ltd. Utilization of Natural Gas in the Ni-
    gerian Economy. Calgary: Government of Canada, External
    Aid Office, May 1964.

Central Bank of Nigeria. Annual Report and Statement of Accounts,
    December 31, 1971.

Curry, Robert L. , and Donald Rothchild. "On Economic Bargaining
    between African Governments and Multinational Companies."
    Journal of Modern African Studies, no. 2 (1974), p. 173.

Dunning, John H. , ed. The Multinational Enterprise. London:
    George Allen and Unwin Ltd. , 1971.

East African Academy. Research Services in East Africa. Nairobi:
    East African Publishing House, 1966.

East African Agricultural and Forestry Research Organization.
    Record of Research Annual Report 1969. Nairobi: East Afri-
    can Community, 1970.

Federal Republic of Nigeria, Second National Development Plan
    1970-74. Lagos: Federal Ministry of Information, 1970.

Frankel, P. H. Essentials of Petroleum: A Key to Oil Economics.
    New York: Augustus M. Kelley Publishers, 1969.

Granick, David. "Economic Development and Productivity Analysis:
    The Case of Soviet Metalworking." Quarterly Journal of Eco-
    nomics, May 1957.

Hartshorn, J. E. Politics and World Oil Economics. New York:
    Praeger, 1967.

Hetman, Francois. Society and the Assessment of Technology.
    Paris: Organization for Economic Cooperation and Develop-
    ment Publications, 1974.

Kahnert, F. et al. Economic Integration Among Developing Coun-
    tries. Paris: Organization for Economic Cooperation and
    Development Publications, 1969.

Katz, Jorge M. "Industrial Growth Royalties Paid Abroad and Local
    Expenditures on R & D." Paper presented to IEA Conference
    on Latin American Development, December 1971, in Mexico
    City. Mimeographed.

Keen, B. A. "The East African Agricultural and Forestry Research
    Organization--Its Origins and Objectives." Nairobi: the East
    African Standard, Ltd., no date.

Lewis, W. Arthur. "Aspects of Industrialization." In 15th Anni-
    versary Commemoration Lectures. Cairo: National Bank of
    Egypt, 1953.

Nelson, Richard. "Less Developed Countries, Technology Transfer
    and Adaptation, and the Role of the National Science Community."
    Yale Economic Growth Center Discussion Paper 104, January
    1971.

Organization for Economic Cooperation and Development. Stock of
    Private Direct Investments by DAC Countries in Developing
    Countries. Paris: OECD Publishers, 1972.

Pack, Howard. "Employment in Kenyan Manufacturing." Yale Eco-
    nomic Growth Center Paper, January 1972.

Pearson, S. R. "Nigerian Petroleum: Implications for Medium-Term
    Planning." In Growth and Development of the Nigerian Economy,
    edited by Carl K. Eicher and Carl Liedholm, pp. 352-75.
    Michigan State University Press, 1970.

_____. Petroleum and the Nigerian Economy. Stanford: Stanford
    University Press, 1970.

Ranis, Gustav. "Relative Prices in Planning for Economic Develop-
    ment." In International Comparisons of Prices and Output,
    edited by D. J. Daly. National Bureau of Economic Research.
    New York: Columbia University Press, 1972.

Report of the Commission on International Development. Lester B.
    Pearson, chairman. Partners in Development. New York:
    Praeger, 1969.

Report of the Commission on the Most Suitable Structure for the
    Management, Direction and Financing of Research on an East
    African Basis. A. C. Fraser, chairman. Nairobi: Govern-
    ment Printer, 1961.

Robson, Peter. Economic Integration in Africa. London: George
    Allen and Unwin Ltd., 1968.

Schatzl, L. H. Petroleum in Nigeria. Oxford University Press for
    the Nigerian Institute of Social and Economic Research, 1969.

Schlie, T. W. "Some Aspects of Regional-National Scientific Rela-
    tionships in East Africa." Ph.D. dissertation, Northwestern
    University, 1973.

Schumacher, E. F. Small is Beautiful: Economics As If People
    Mattered. New York: Harper and Row Publishers, 1974.

Singer, H. W. "The Distribution of Gains Between Investing and
    Borrowing Countries." American Economic Review, 1950,
    pp. 473-85.

Singh, K. D. N. UNIDO Guidelines for the Acquisition of Foreign
    Technology in Developing Countries (with special reference to
    technology license agreements), U.N. ID/98, New York, 1973.

Smith, Edward H. "Transfer of Technology, Choice of Techniques
    and Economic Growth." Ph.D. dissertation, Yale University,
    1972 or 1973.

Stobaugh, R. B. The International Transfer of Technology in the
    Establishment of the Petrochemical Industry in Developing
    Countries, UNITAR Research Report No. 12, UNITAR, New
    York, 1971.

Tanzer, Michael. The Political Economy of International Oil and the
    Underdeveloped Countries. Boston: Beacon Press, 1969.

Thomas, D. B. Capital Accumulation and Technology Transfer:
    Some Theoretical and Empirical Analyses of the Nigerian
    Manufacturing Industries. New York: Praeger, 1975.

_____. "La transferencia internacional de technologia industrial y las naciones nuevas." El Trimestre Economico 41(3), no. 163. Mexico: July 1974.

Turman, David. The Employment Problem in Less Developed Countries. Paris: OECD Development Centre, 1971.

United Nations. Department of Economic and Social Affairs. Multinational Corporations in World Development. ST/ECA/190, 1973.

_____. The Impact of Multinational Corporations on Development and on International Relations. New York: United Nations, 1974.

_____. Summary of the Hearings Before the Group of Eminent Persons to Study the Impact of Multinational Corporations on Development and on International Relations. ST/ESA/15, 1974.

Worthington, E. B. A Survey of Research and Scientific Services in East Africa, 1947-1956. Nairobi: East African High Commission, 1952.

ABOUT THE AUTHOR
AND CONTRIBUTORS

D. BABATUNDE THOMAS is associate professor of economics
and technology and chairman of the Economics Department at Florida
International University, Miami, Florida. He was born in Nigeria
and educated in Nigeria, Great Britain, and the United States. Dr.
Thomas received his Ph.D. in economics from Indiana University.
He specialized in economic development and planning, econometrics,
and international business administration. He is a frequent contribu-
tor to international conferences on technology and economic development, a consultant, a contributing editor, and the author of several
articles on the subject. He also is the author of Capital Accumulation and Technology Transfer: A Comparative Analysis of Nigerian
Manufacturing Industries (New York: Praeger, 1975). Dr. Thomas
has taught at Indiana University and Kalamazoo College. Before going
to Florida International University, he was a research fellow at the
Nigerian Institute of Social and Economic Research at the University
of Ibadan, Nigeria.

MIRA WILKINS is professor of economics at Florida International University. Dr. Wilkins is the author of numerous articles in
scholarly journals and a two-volume history of multinational corporations, The Maturing of Multinational Enterprise: American Business
Abroad from 1914 to 1974 (Cambridge, Mass.: Harvard University
Press, 1974) and The Emergence of Multinational Enterprise: American Business Abroad from the Colonial Era to 1914 (Cambridge,
Mass.: Harvard University Press, 1970).

WALTER A. CHUDSON is advisor of foreign investment, Department of Economic Affairs, United Nations, N.Y. He is author
of The International Transfer of Commercial Technology to Developing Countries (New York: United Nations Institute for Training and
Research, 1971). Dr. Chudson's views expressed in Chapter 5 of
this book are not the official views of the United Nations or any of
its agencies. An earlier version of his chapter was presented to the
panel on "Barriers to Effective International 'Transfer' and Diffusion
of Technology to African Countries." The panel was chaired by Dr.
Thomas in October 1973 at the African Studies Association International Conference in Syracuse, New York.

THEODORE W. SCHLIE is a member of the research staff at
Denver Research Institute. Dr. Schlie also delivered an earlier version of his chapter to the panel chaired by Dr. Thomas in Syracuse
in October 1973.

RELATED TITLES
Published by
Praeger Special Studies

CAPITAL ACCUMULATION AND TECHNOLOGICAL
TRANSFER: A Comparative Analysis of Nigerian
Manufacturing Industries
D. Babatunde Thomas

LEGAL ASPECTS OF THE INTERNATIONAL
TRANSFER OF TECHNOLOGY TO DEVELOPING
COUNTRIES
Charles C. Okolie

ADMINISTRATIVE TRAINING AND DEVELOPMENT:
A Comparative Study of East Africa, Zambia,
Pakistan, and India
Bernard Schaffer

*TECHNOLOGY IN COMECON: Acceleration of
Technological Progress through Economic Planning
and the Market
J. Wilczynski

---

*For sale in the United States and Philippines only.